BUCKNELL REVIEW

Adrift in the
Technological Matrix

STATEMENT OF POLICY

  BUCKNELL REVIEW is a scholarly interdisciplinary journal. Each issue is devoted to a major theme or movement in the humanities or sciences, or to two or three closely related topics. The editors invite heterodox, orthodox, and speculative ideas and welcome manuscripts from any enterprising scholar in the humanities and sciences.

This journal is a member of the Conference of Editors of Learned Journals

BUCKNELL REVIEW
A Scholarly Journal of Letters, Arts, and Sciences

Editor
GREG CLINGHAM

Associate Editor
DOROTHY L. BAUMWOLL

Assistant Editor
ANDREW P. CIOTOLA

Contributors should send manuscripts with a self-addressed stamped envelope to the Editor, *Bucknell Review,* Bucknell University, Lewisburg, PA 17837.

BUCKNELL REVIEW

Adrift in the Technological Matrix

Edited by
DAVID L. ERBEN

Lewisburg
Bucknell University Press
London: Associated University Press

Associated University Presses
2010 Eastpark Boulevard
Cranbury, NJ 08512

Associated University Presses
16 Barter Street
London WC1A 2AH, England

Associated University Presses
P.O. Box 338, Port Credit
Mississauga, Ontario
Canada L5G 4L8

The paper used in this publication meets the
requirements of the American National Standard for
Permanence of Paper for Printed Library Materials Z39.48-1984.

(Volume XLVI, Number 2)

ISBN 0-8387-5551-8
ISSN 0007-2869

PRINTED IN THE UNITED STATES OF AMERICA

Contents

Recent Issues of BUCKNELL REVIEW

Classics and Cinema
Reconfiguring the Renaissance: Essays in Critical Materialism
Wordsworth in Context
Turning the Century: Feminist Theory in the 1990s
Black/White Writing: Essays on South African Literature
Worldviews and Ecology
Irishness and (Post)Modernism
Anthropology and the German Enlightenment: Perspectives on Humanity
Having Our Way: Women Rewriting Tradition in Twentieth-Century America
Self-Conscious Art: A Tribute to John W. Kronik
Sound and Light: La Monte Young/Marian Zazeela
Perspectives on Contemporary Spanish American Theatre
Reviewing Orpheus: Essays on the Cinema and Art of Jean Cocteau
Questioning History: The Postmodern Turn to the Eighteenth Century
Making History: Textuality and the Forms of Eighteenth-Century Culture
History and Memory: Suffering and Art
Rites of Passage in Ancient Greece: Literature, Religion, Society
Bakhtin and the Nation
New Essays in Ecofeminist Literary Criticism
Caribbean Cultural Identities
Lorca, Buñuel, Dalí: Art and Theory
Untrodden Regions of the Mind: Romanticism and Psychoanalysis
Art and the Religious Impulse

7

Acknowledgments

The idea for this collection grew out of a series of virtual "events" in the early 1990s organized around the topic of the new communications technologies held at the University of South Florida. The events consisted of a series of email and MUD exchanges and a conference, "King Ludd and the Resistance to Technology." Influential texts for these events include the marvelous *High Techne: Technology and Art in Modernity and Beyond* by R. L. Rutsky and *Virtual Realities and Their Discontents*, edited by Robert Markley. I would also like to thank the following institutions and individuals for their help during the writing and editing of *Adrift in the Technological Matrix:* Kelli Sebastian, Phil Sipiora, John Unsworth, John C. Rowe, Mark Poster, N. Katherine Hayles, Kali Tal, the University of South Florida, and the Institute for Advanced Teaching in the Humanities at the University of Virginia. Special thanks go to Robert Markley and Mark C. Taylor, without whom this collection would never have made the light of day.

Notes on Contributors

GEOFFREY BENNINGTON is Asa G. Chandler Professor of Modern French Thought at Emory University. His two most recent books are *Interrupting Derrida* and *Frontières kantiennes*.

J. YELLOWLEES DOUGLAS is an associate professor of English at the University of Florida. She has published extensively on the interface between cognition, reading, representation, interactive media, and product design. Most recently, she has published articles on distance education, on the interface between industrial design, interaction, and user schemas, and on immersion and engagement in interactive narratives.

JOHANNA DRUCKER is a professor of English and Director of Media Studies at the University of Virginia. Her scholarly books include *Theorizing Modernism; The Visible Word: Experimental Typography and Modern Art; The Alphabetic Labyrinth; The Century of Artists' Books;* and, her most recent collection, *Figuring the Word*. Drucker is internationally known as a book artist and experimental, visual poet. Recent titles in this area are *Prove Before Laying; Night Crawlers on the Web; Nova Reperta;* and *Emerging Sentience,* the last two in collaboration with Brad Freeman. *A Girl's Life,* a collaboration with painter Susan Bee, is forthcoming.

DAVID L. ERBEN is an assistant professor at the University of Toledo and of Mescalero descent. He became interested in the Internet and the new computer technologies in the early 1990s when they were promoted as emancipatory and revolutionary technologies, particularly for marginalized and minority peoples in the U.S. and Eastern Europe. Currently, his work on technology focuses on the way it has failed to live up to these expectations.

SILVIO GAGGI is Chair of the Department of Humanities and American Studies at the University of South Florida, where he teaches interdisciplinary courses focusing on aspects of twentieth-century culture and critical theory. He has published articles dealing with modern art, literature, film, and theater, as well as two books, *Modern/Postmodern: A Study in Twentieth-Century Arts and Ideas*, and *From Text to Hypertext: Decentering the Subject in Fiction, Film, the Visual Arts, and Electronic Media*. He is currently involved in research related to identity, the body, and new technologies.

CATHERINE GOUGE is a visiting assistant professor of English at West Virginia University where she has been designing and piloting Web-based courses for the English Department. She is currently working on projects that examine the role of frontierist ideology in American culture and which explore how a frontierist American culture is constructed through and by narratives of technology and citizenship.

DAVID KOLB is Charles A. Dana Professor of Philosophy at Bates College. His writings concern issues about the nature of modernity and postmodernity, as found in nineteenth- and twentieth-century German philosophy, architecture and planning, and hypertext and other new media of communication. His publications include *The Critique of Pure Modernity: Hegel, Heidegger, and After; Postmodern Sophistications: Philosophy, Architecture, and Tradition;* and *Socrates in the Labyrinth: Hypertext, Argument, Philosophy*. Currently he is working on "Sprawling Places," a discussion of the concept of place, with applications to supposed "nonplaces" such as suburban sprawl and theme parks.

ROBERT MARKLEY is Jackson Distinguished Chair of British Literature at West Virginia University. He has published numerous articles on seventeenth- and eighteenth-century literature, cultural studies, and the relations between literature and science. His most recent books include *Fallen Languages: Crises of Representation in Newtonian England, 1660–1740; Virtual Realities and Their Discontents* (editor). Two forthcoming titles are *Dying Planet: Mars and the Anxieties of Ecology from the Canals to Terraformation* and *Fictions of Eurocentrism: The Far East and the English Imagination, 1600–1800*. He is editor of *The Eighteenth Century: Theory and Interpretation*, The Series on Science and Culture for the University of Oklahoma Press, and co-editor of

Mariner 10: Educational Multimedia for the University of Pennsylvania Press. He is co-author of *Red Planet: Scientific and Cultural Encounters with Mars,* the first scholarly title authored for DVD-ROM.

MARK C. TAYLOR is Cluett Professor of Humanities at Williams College and co-founder of the Global Education Network. His most recent book is *The Moment of Complexity: Emerging Network Culture and Grave Matters,* which also will be the subject of an exhibition at Mass MOCA in the fall of 2002.

Introduction: Freedom's Absent Horizon in the Technological World

TECHNOLOGY has consistently transformed the world by a process that seems to be constantly accelerating. Yet, the process of technological innovation is all too often discussed as though it were occult and inevitable.[1] The struggle to understand the way the "new" computer and communications technologies is transforming the world is definitely complicated. What the essays collected in this issue of the *Bucknell Review* attempt is a general cultural approach to the notion of there being a technological "matrix" in which we all now find ourselves "adrift" and of which our experience is often "dread." "Adrift" and "dread" are not single metaphors in the collection.[2] The metaphors are allowed to transform themselves from essay to essay. In order to interrogate the technological matrix, the authors have drawn from a variety of disciplines—literature, philosophy, religion, art, media studies—while retaining the great substantial contributions of previous theorists of technology. The main thrust of this collection is to underscore the vast enrichment given to a study of the "new" technologies when approached from a cultural standpoint, i.e., to show the advantages of such an approach over the isolationist tendencies of most pure scientific studies that have the characteristic of being highly abstract and unintelligible.

Throughout these essays, by means of various readings of the technological space we find ourselves in, by academic scholars and practicing artists, the authors demonstrate the remarkable cultural cross-fertilization among the disciplines in regard to the "new" technologies. The ways in which philosophical, artistic, and scientific concepts and discoveries have metaphorically penetrated our relationship with technology is one of the threads that weaves the essays together. Following Heidegger, the authors all in one way or another address the fundamental question: is technology something we do or is it something done to us?[3]

To posit the notion of being adrift within a "technological ma-

15

trix" is to metaphorically create a threshold space, a frame. From the Oxford English Dictionary we discover that "matrix" provides a number of appropriate points of departure. There is the sense of a matrix being a "place of birth where something is bred or produced." It is also a place where things are "connected" in some way and something which also "surrounds." So the metaphor has spatial, originary, and associative senses. This issue of the *Bucknell Review* is concerned with a number of broad issues in this matrix, including the political and epistemological relationships between theory and technology. But this frame, this matrix, itself is overlaid with, and becomes an allegory of, is inscribed into, a whole range of issues "unpacked" in the essays which follow.

We are living in a world overdetermined by technology, by a technological "matrix," consisting of cyberspace, cellular phones, handhelds, DVD and other technical resources and the accompanying language ("ripping," "hacking," "surfing," "Web," "wired," etc.): a matrix of electronic, linguistic and media networks and rhizomatic relations. However the nature and effects of this matrix and these relations still largely remain unclear. Frequently associated with postmodern movements like cyberpunk fiction and music and extreme individualism (for example "computer hacking"), this matrix often suggests utopian visions of collectivity, role-playing and of "free and unmediated" communications exchanges using virtual environments. Lumped together with emancipatory political and social movements in Eastern Europe, Africa, Asia, and Central and South America that have taken advantage of the portability of the new communication technologies (for example the Zapatista movement's use of real-time video-broadcasting via the Internet to call attention to their very "real" resistance in Chiapas) and the potential offered by the World Wide Web for disseminating information, this technological matrix contains both the possibility of being just another cliché that is distracting us from the social and economic changes that have been occurring in the postmodern era and the possibility of a profound revolutionary technology. And, of course, it is also possible to read the experience of this technology in terms of "dread" "and drift," in terms of our sense of being thrown into a technological world not of our own making and of which we have little control.

It is also very important to keep in mind in regard to this matrix that it is also at the heart of a postmodern economy. This economy relies primarily on the exchange of representations and information. All material production for this matrix is subordinate to the discursive practices and representational logics of this virtual and postmodern economy. Postmodern subjects have not been "liberated" from corporality in this economy—Nietzsche's time, change, and becoming; corporality is just simply subordinated to an adaptable system of production and exchange. More precisely, physical and biological organisms and processes are being transformed into information processes, thus complicating the way we think of the "body."

To ask what a thing is, is to ask about its essence. There are ready answers to the question, What is technology? in which the essence takes the form of some kind of definition. For example, technology defined as a means to an end contains two related positions. It points to all the devices and equipment that "do things" for us. In this sense a computer is just a "means" for accomplishing a task. However, technology becomes problematic when these various means have undesirable consequences of one sort or another so that the major issue becomes finding the best and most effective means for whatever ends we have in mind.

The issue of the politics of the matrix is also very complex. It is precisely the point where some of the political debates surrounding postmodernism and poststructuralism become relevant. Part of what is involved with this technological matrix is a reconfiguring of the infrastructure/superstructure political model. The question of the politics of the matrix may ultimately be a simple one: not what politically is within the technological matrix, but what is not within it, what is not contaminated by it.

Historically, there has always been disagreement over the nature of technology and of course there has been a recent explosion of books and articles in mainstream magazines like *Time* and *Newsweek*, scholarly journals, and both popular and scholarly books. Surveying the now seemingly quite distant past, however, to get a glimpse of what came before, Ortega y Gasset conceived of technology as an expression of human nature. For Teilhard de Chardin, technology is an expression of human ends. Buckminster Fuller thought that technological development had not gone far enough. Barry Commoner argues that technology and science are out of control. Theorists like E. F. Schumacher and Herbert Marcuse stress the importance of technology as a certain kind of thinking. Two of the most influential contemporary philosophers of technology and modernity are Jurgen Habermas and Martin Heidegger. For Heidegger, technology is to be understood as a particular way that the things of the world come to be present and for Habermas the dominant pathology of modernity is the colonization of life by technology.

Adrift in the Technological Matrix offers a broad, provocative and often playful critical approach to the resources we associate with the technological matrix that surrounds and influences us, investigating the relation of embodied practices to reading and writing—the kinesthetic, tactile, and visual, and how these affect the experience of this space. We begin with a general introduction by Mark C. Taylor.

DAVID L. ERBEN

Notes

1. Note the invention of the telephone and Watson's hope that it would enable communication with spirits.

2. I am following here the sense of "dread" developed by Heidegger: the sense of terror produced by being in the presence of complete otherness.

3. Martin Heidegger, *The Question Concerning Technology and Other Essays*, ed. and trans. William Lovitt (New York: Harper & Row, 1977).

BUCKNELL REVIEW

Adrift in the Technological Matrix

cult@edu.com

Mark C. Taylor
Williams College

T HINGS are shifting and we are caught in a middle that seems to have neither a clear beginning nor evident ending. This situation is not, of course, new, for we are always betwixt and between without any certainty of where we have come from and where we are heading. Nevertheless, this time things appear to be different—the shift seems more shifty, the middle more muddled. There is a pervasive awareness that something is passing away but no clear sense of what is emerging. Many people realize that we live in a "post" age but have no idea how to imagine what might be coming next. Indeed, the post age is largely defined by a preoccupation with recycling the old and suspicion of the new. When everything seems caught in the web of the "always already," we are left with a recombinant culture in which creative innovation no longer seems possible. Been there, done that—always already been there, done that.

And yet, things *are* shifting—something *is* passing away but something *is* also emerging. Not everyone shares the suspicion of the new; indeed, now, as always, prophets confidently proclaim the New Age that is arriving in our midst. The louder prophets preach, however, the more suspicious critics become, and, conversely, the more critics criticize, the louder the prophets shout. This struggle between prophets and critics involves contrasting assessments of the impact of technology on social, political, economic, and cultural processes. As the millennium begins, many technophiles are rewriting the Western metanarrative of progress to fit the information age.

To appreciate the importance of this shift, it is helpful to understand its historical context. While critics often interpret the excessive faith in technology as an extension of the Enlightenment narrative of the liberating effects of universal reason, the utopian expectations of technophilics actually must be traced to the notion of salvation history, which characterizes the Western theological tra-

21

dition. In the secular version of this three-part story, the ages of the
Father, Son, and Spirit become the eras of agrarianism (farming),
industrialism (manufacturing), and postindustrialism (information).[1] Throughout the course of modernity, belief in the salvific potential of technology has been widespread and sometimes is found
among unlikely suspects. During the early decades of this century,
art effectively displaced religion as the vehicle for human salvation.
Influential members of the avant-garde insisted that the artist is a
modern prophet who would lead the way to the Promised Land,
which was envisioned as a utopian human community. Such a community would, in effect, be a beautiful work of art. No longer iso-
lated from the world by being hung on a wall or set on a pedestal,
art becomes worldly as the world is transformed into a work of art.
This transformation is the *end* of the work of art in every sense of the
term. On the occasion of the controversial 1921 exhibition entitled
"5 × 5 = 25" mounted by Moscow's Institute for Artistic Culture
(INKhUK), Rodchenko echoed Hegel and Nietzsche's proclama-
tion of the death of God by confidently declaring:

> Art is dead! . . . Art is as dangerous as religion as an escapist activity . . .
> Let us cease our speculative activity and take over the healthy bases of
> art—color, line, materials and forms into the field of reality, of practical
> construction.[2]

However, just as the death of God is not a simple negation but is a
complex process in which the divine becomes incarnate when the
profane is grasped as sacred, so art ends not because it disappears
but because it appears everywhere. Art ends when everyone, in Andy
Warhol's famous words, "becomes an artist" and the world itself finally becomes a work of art. For Rodchenko, the move from "speculative activity" to "practical construction" entails a commitment to
create socially useful products. Turning from "pure art" to "production art," he moved from gallery to factory and began to create pro-
paganda posters and industrial designs for state industries.

As the twentieth century unfolded, a darker side to this wedding
of modernism and modernization gradually appeared. When artists
of vastly different political persuasions sought to make their work
productive through mechanization and industrialization, art gradually lost its critical edge and became an instrument by means of
which the military-industrial complex realized its ends. Gertrude
Stein, whose portrait Picasso so carefully painted, went so far as to
describe World War I as a "cubist war":

Really the composition of this war, 1914–1918, was not the composition of all previous wars, the composition was not a composition in which there was one man in the center surrounded by a lot of other men but a composition that had neither a beginning nor an end, a composition of which one corner was as important as another corner, in fact the composition of cubism.[3]

The war, however, was not only cubist but was perhaps more importantly futurist. Conflict on a global scale presupposed an extraordinary increase in speed that manifested itself in all dimensions of life. Stephen Kern points out that the crisis of August 1914 would have been "unfathomable to anyone who had lived before the age of electronic communication. In the summer of 1914 the men in power lost their bearings in the hectic rush paced by flurries of telegrams, telephone conversations, memos, and press releases; hard-boiled politicians broke down and seasoned negotiators cracked under the pressure of tense confrontations and sleepless nights, agonizing over the probable disastrous consequences of their snap judgments and hasty actions."[4] Anticipating what many today regard as dangers of the information age, Kern leaves little doubt about the devastating consequences wrought by "advances" in technology. But not even the horrors of war could dampen the technological enthusiasm of a futurist like Filippo Marinetti, who preached a gospel of techno-futurism in which the devastation of war was a "purifying ritual" that would lead to a brighter future:

> We affirm that the world's magnificence has been enriched by a new beauty; the beauty of speed. A racing car whose hood is adorned by great pipes, like serpents of explosive breath—a roaring car that seems to run on shrapnel—is more beautiful than the Victory of Samothrace.
> We glorify war—the world's only hygiene . . . We will sing of great crowds excited by work, by pleasure, and by riot; we will sing of the multi-colored, polyphonic tides of revolution in the modern capitals.[5]

While Marinetti's disturbing declaration is excessive, it is far from unique and echoes in the voices and works of all too many modernists. Le Corbusier, whose dreams of a rational utopia are well known, celebrates what he calls "the 'White World'—the domain of clarity and precision, of exact proportion and precise materials, culture standing alone—in contrast to the 'Brown World' of muddle, clutter, and compromise, the architecture of inattentive experience."[6] These racist words become even more ominous when one notes the inscription above the title on the first page of *The Radiant City*: "THIS WORK IS DEDICATED TO AUTHORITY, PARIS, MAY, 1933." A year earlier, Le Corbusier had become an editor of the journal *Prelude,* whose board included several well-known pro-Fascists. Though Le Corbusier was initially critical of fascism, he quickly changed his mind when Mussolini invited him to Italy to explore the possibility of designing buildings for his regime. Shortly thereafter, Le Corbusier wrote in Marinetti's pro-Fascist publication, *Stile futurista*: "The present spectacle of Italy, the state of her spiritual powers, announces the imminent dawn of the modern spirit. Her shifting purity and force illuminate the paths which had been obscured by the cowardly and the profiteers."[7] While the rhetoric of spiritualism and utopianism remains, the substance of the vision has changed considerably. It should not be surprising that the unholy alliance of modernism, industrialism, and militarism created widespread suspicions about the impact of technology.

For many insightful critics, the relation between industrialism and the information age is marked more by continuity than discontinuity. The information era has been brought about by technological innovations that were first designed and implemented for military purposes. From cathode tubes and radar screens to computers and Internet, the driving force behind research and development has been national defense. While the military-industrial complex surely has not disappeared, a new military-information complex has emerged. As technology has become more sophisticated, the weapons and the networks supporting them have become smarter. During the half-century after the end of World War II, government and business used the ideology of the Cold War to justify vast expenditures for information and telematic technologies. With the collapse of communism and the end of the Cold War, new narratives had to be constructed to explain developments and justify what was increasingly touted as the information revolution. As in the past, technology designed for military purposes was rapidly adapted and diverted to commercial ends. For government, business, and consumers, this seemed to be a win-win-win strategy.

The obvious continuities between industrialism and postindustrialism notwithstanding, there are no less significant differences between them. The shift from a manufacturing to an information economy is inseparable from fundamental changes in the means of production and, more important, reproduction in the capitalist economy. When information technologies began to find applications beyond the military sector, the shape of capitalism began to change. The transformation of capital into information and vice versa combined with the rapid spread of telematic technologies to create a consumer capitalism in which information and media joined in positive feedback loops that resulted in exponential growth. The collapse of the Berlin Wall arguably had more to do with the lure of consumer capitalism generated by burgeoning media networks than with carefully crafted political and economic strategies. Once again it is possible to interpret these developments as the latest version of the ancient story of salvation history. Consumer culture, its promoters would have us believe, is the realized kingdom or worldly utopia that humankind has long awaited.

Though not immediately apparent, there again is a curious alliance between art and technology in the rise of information and consumer capitalism. What began in Rodchenko's Russia came to an unanticipated end in Warhol's United States. While Rodchenko took his artistic skills into the factory to produce advertising and products that promoted world socialism, Warhol brought ads and commodities into his studio, which he dubbed The Factory, to produce images that promoted global consumer capitalism. In Warhol's Factory, art became business and business became an art. The work of art in the age of electronic reproduction holds up a mirror in which the world sees itself reflected. The business of art is the inverted image of the art of business in a consumer culture where the consummate art is the art of the deal. Warhol's art makes money— both as product and as process. Unashamedly confessing in words that are not merely ironic, he declares:

> I don't understand anything except GREEN BILLS. Business art is the step that comes after Art. I started as a commercial artist, and I want to finish as a business artist. After I did the thing called "art" or whatever it's called I went into business art. I wanted to be an Art Businessman or a Business Artist. Being good in business is the most fascinating kind of art. During the hippie era people put down the idea of business—they'd say, "Money is bad," and "Working is bad," but making money is art and working is art and good business is the best art.[8]

When the religious narrative of the dawning of the New Age is rewritten in artistic terms, its conclusion seems to be either totalitarianism or fascism on the one hand, or consumer culture and global capitalism on the other. Rather than the realization of the kingdom of God on earth, heaven seems to have become hell through a reversal that must itself be reversed.

Opposites often have a way of being more similar than they initially appear. While Soviet socialism and consumer capitalism are inspired by antithetical ideologies, a growing number of critics realize that they share certain problematic characteristics. On the most general level, they both can be characterized by what might be described as a "totalizing logic." On the one hand, centralized government and authoritarian power and, on the other, the hegemonic machinations of capital create systems and structures that tend to become all consuming and thus totalistic. Some critics detect a similar logic at work in the information and media networks that are rapidly proliferating. What Jameson and others define as postindustrial or multinational capitalism might better be described as information capitalism, which is, of course, global. In his provocative essay, "Eclipse of the Spectacle," Jonathan Crary draws on the work of Foucault to argue that "the disciplinary apparatus of digital culture poses as a self-sufficient, self-enclosed structure without avenues of escape, with no outside."[9] What seems to make digital culture so irresistible is the seeming "hegemonic" power of the code. With the development of digital technologies that transform information and media into code, what writers from Aristotle to Marx identify as the universal or abstract equivalent expands from the economic sphere to encompass all aspects of culture. As the power of the code spreads, exchange value assimilates use value. Crary underscores the significance of these developments in a suggestive comment on Pynchon's *Gravity's Rainbow.*

> It is a transition coinciding with the processes of rationalization that Pynchon describes, the abstract coding of anything that would claim singularity, and also with television annihilation of the "semantic field" in Ballard. What *Gravity's Rainbow* tells us better than any other text is how World War II was above all an operation of modernization: how it was the necessary crucible for the obliteration of outdated territories, languages, filiations, of any boundaries or forms that impeded the installation of cybernetics as the model for the remaking of the world as pure instrumentality. And it cannot be overemphasized how the development of cybernetics ("a theory of messages and their control") is intertwined with the commodification of all information and with the hegemony of what Pynchon calls the "meta-cartel."[10]

When code becomes capital and capital becomes code, the processes of commodification seem to know no bounds. The "meta-cartel" is not any specific business or industry but a worldwide web that creates a postindustrial global information-entertainment complex. Faced with the accelerating expansion of global capitalism facilitated by the explosive growth of information and telecommunications networks, critics desperately seek to develop strategies of resistance to hinder if not subvert what they regard as repressive policies and practices. Crary proposes a tactic that by now has become familiar. To resist the growing domination of the abstract code, he maintains, we must find gaps in the web by returning to the irreducible materiality of the body and its deliberate rhythms:

> Perhaps the most fragile component of this future, however, has been in the immediate vicinity of the terminal screen. We must recognize the fundamental incapacity of capitalism even to rationalize the circuit between body and computer keyboard, and realize that this circuit is the site of a latent but potentially volatile disequilibrium . . . Its [digital culture's] myths of necessity, ubiquity, efficiency, of instanteity require dismantling: in part, by disrupting the separation of cellularity, by refusing productivist injunctions, by inducing slow speeds and inhabiting silences.[11]

The preoccupation with the body in much recent art and criticism is, at least in part, an expression of the effort to identify a site of resistance to the processes of dematerialization and virtualization characteristic of contemporary experience. As "the society of the spectacle" gives way to the virtual realities of digital culture, ancient dreams of otherworldly escape reappear in what Eric Davis dubs "techgnosis."[12] Signaling the dangers inherent in a *"systematic devaluation of materiality and embodiment,"* Katherine Hayles cautions:

> Information, like humanity, cannot exist apart from the embodiment that brings it into being as a material entity in the world; and embodiment is always instantiated, local, and specific. Embodiment can be destroyed, but it cannot be replicated. Once the specific form constituting it is gone, no amount of massaging data will bring it back. This observation is as true of the planet as it is of an individual life form. As we rush to explore the new vistas that cyberspace has made available for colonization, let us remember the fragility of a material world that cannot be replaced.[13]

There are many ways to respond to the challenge Hayles so effectively poses. While the interests and preoccupations of critics vary, much of their debate during the past thirty years bears directly and indirectly on questions raised by the intersection of informatics, telematics, and consumer capitalism. Critics who agree about little else join in condemning the deleterious effects wrought by systems and structures that tend to repress difference and otherness. Whether framed in terms of social, political, and economic hegemony or linguistic, literary, and philosophical totalization, critical moves routinely involve soliciting the repressed by revalorizing the marginalized or excluded. As I have noted, when faced with new technologies and the changes they bring, analysts turn to critical perspectives that seem to have been effective in the past. Some extend the work of members of the Frankfurt School like Adorno and Horkheimer to reformulate a critique of the culture industry and its instrumental reason in the digital age. Others, like Crary, appropriate Foucault's genealogical analyses to elaborate a politics that shifts attention from the global and macro to the local and micro. For critics who are more concerned with the psychological than the sociopolitical aspects of the problem, Lacan's rereading of Freud provides useful resources. Still others deploy deconstructive criticism to detect fissures in the apparently seamless edifice of information capitalism. In addition to these more or less philosophical approaches, many critics have been concerned about the social, political, and economic impact of emerging informatics and cyberworlds. Drawing on several of these strands, feminists, for example, examine the ways in which new technologies reinforce old gender roles. Running through all of these analyses and criticisms there is a growing recognition that when information becomes the currency of the realm, power is redistributed in significant ways. Questions of access to and control over technology are becoming more important for all aspects of human life. It is precisely this recognition that lends the critiques of informatics and telematics their urgency.

Without denying the usefulness of many of these criticisms, it is necessary to admit that they are quite predictable. Arguments that once seemed innovative have become tired and worn and, thus, too often seem utterly formulaic. And yet, the more frayed the arguments become, the more insistently they are repeated. Of what is this compulsion to repeat a symptom? Why are so many people in the university so unnerved by and resistant to information and telecommunications technologies?

There are no simple answers to these questions; nevertheless, some of the reasons for the pervasive suspicions about digital culture are beginning to become clear. While usually cast in terms of economic class, social privilege, and political power, what actually seems the pivotal issue for many critics is the growing threat that information technologies undeniably pose to their own interests. Beneath the seemingly high-minded appeals on behalf of the disempowered and underprivileged, it is possible to detect a strong—sometimes reactionary—defense of personal and institutional power and privilege. In some ways, the so-called information revolution extends the industrial revolution by expanding processes of commodification to the sphere of culture. But this apparent continuity actually constitutes a significant difference between industrial and information capitalism, which poses a significant threat to the university in its current form. As education and culture are commodified, the university loses its effective monopoly on the means of the production and reproduction of knowledge and increasingly becomes subject to market forces. Critics on the left and the right are united in their opposition to information capitalism. The left is critical of the commodification of culture and education, which extends the totalizing logic and hegemonic power of capital, and the right is distressed by the invasion of the university by market forces that subvert the purported "intrinsic value" of education. From both points of view, the source of corruption, it seems, is filthy lucre. Warning of the dire consequences that will result from the extensive deployment of information technologies in higher education, sociologist Robert Bellah argues that the "coercion of the market" forces the university into the untenable position of "justifying itself in terms of its contribution to external ends."[14] But Bellah's values are not as "intrinsic" as he insists; indeed, his analysis suggests that he shares the self-interested values of the market he claims to resist:

> The tyranny of the bottom line drives academic decisions in several ways. When the university is seen simply as part of the economy, then the normal pressures for market efficiency set in, and the consequences are nowhere more ominous than in the sphere of personnel decisions. Contemporary industry wants to control labor costs, and downsizing is a common mechanism for doing so. It is difficult to cut the number of instructors, since a certain number of classes must be taught, and in public universities rising enrollments is creating pressure for more classes. Nevertheless, some colleges and universities have resorted to simple downsizing by cutting faculty, expanding the teaching load, and increasing class size.[15]

What Bellah does not state explicitly in this context but repeatedly implies elsewhere is that many administrators are attempting to use information technologies to increase efficiency and thereby reduce the number of faculty members required to "deliver the product." The issue, then, is not that the intrinsic value of education and culture is threatened by the market but that job security is—or, more precisely, job security for those who already have jobs. The real threat that many in the university see in information and telecommunications technology is unemployment or underemployment. Faced with this prospect, they believe it is necessary to resist informatics and telematics in every way possible.

Such resistance creates both theoretical and practical difficulties. In practical terms, the preoccupation with the negative impact of information technologies obscures the creative and constructive potential they harbor. David Kolb suggests, in his essay in this volume, that patient engagement rather than premature dismissal discloses that new immersive media and devices, which deploy digital technology, create "new practices and possibilities" and even "new ways of being." Drawing on traditions as different as Hegelianism and Buddhism, Kolb concludes:

> The modes I have described do not do away with the need for distanced argumentative criticism. Political and cultural criticism may need to back off from immersive artifacts and reformulate them within larger narratives and critical vocabularies. But that is not the only way. Meaning cannot be controlled; it opens new possibilities inside the immersive artifact. I have suggested modes that invoke this openness of meaning and the temporality of our experience. They allow the creativity of impertinent moves. The critical sensitivities involve more than sets of propositions and their inferential connections. We will be impoverished if we envision all criticism on the model of a logical argument, or on the model of a judicial examination.

Kolb's call for the exploration of the creative potential of emerging technologies obviously does not deny the need for sustained critical reflection. To the contrary, a critical distance that nonetheless allows functional appropriation can be maintained. The challenge is to avoid extremes which would either demonize or deify technology. If one is able to walk the fine line between these extremes, it becomes clear that there are also problems with the theoretical perspectives informing the practices of the resistance to and rejection of digital culture.

As I have suggested, critics faced with rapidly changing social, political, economic, and cultural conditions understandably revert to
familiar theoretical perspectives. When this strategy is followed, digital culture tends to be interpreted as the third moment in a narrative whose first two chapters tell the tale of dialectical systems and
binary structures. As Robert Markley points out in his contribution
to this volume, this approach "reinforces the oppositional structures
of thought." More specifically, analyses are framed in terms of traditional oppositions like spirit/body, mind/brain, mechanism/organ-
ism, information/energy, and immateriality/materiality. Whether
conceived as dialectical, binary, or digital, such oppositions are interactions of the foundational contrast between form and matter. As
deconstruction has taught us, oppositions are never equivalent but
are always hierarchical. Throughout the Western tradition, the former term in each of the above oppositions is privileged over the latter. The first move in undoing the logic of oppression is to revalorize
the suppressed term. While necessary, this critical gesture is not sufficient. In other words, it is not enough to attempt to return to the
body or to try to recover material conditions in all of their complexity. Rather, what must be done is to reconceive the matrix in which
the opposites that have long structured our thoughts and grounded
our actions are refigured without being dissolved. When this is
done, form is no longer as formal and matter is no longer as material as they once seemed to be. As this foundational opposition is
recast, the entire structure it grounds shifts.

The enormity of the theoretical and practical stakes of this problem makes the appearance of *Adrift in the Technological Matrix* particularly timely. Caught betwixt and between opposites that once
seemed stable, the condition of drifting provides ever shifting positions from which to assess and reassess changes whose speed we can
hardly fathom. Though we cannot be sure what is emerging, we can
be sure that it will be different from what we have formerly known
and will require us to be different from what we have previously
been. The essays gathered in this issue of *Bucknell Review* approach
the problems and possibilities of our contemporary condition from
productively divergent perspectives. The aim of the inquiry in which
the contributors invite readers to participate is nothing less than the
rethinking of what it means to be human in the dawning "posthuman" age. We ignore this invitation at our own peril.

Notes

1. Alvin Toffler has developed the most influential version of this narrative in his popular book *The Third Wave*. In recent years, Toffler's vision has had a significant political impact. Having become a true believer in Toffler's combination of technofuturism and New Age spirituality, Newt Gingrich asked Toffler to write an introduction to his book on the Republicans' "Contract with America." It is important to note that secularization of biblical narrative can be appropriated to support different political agendas. Leftist critics often turn the story of salvation through technology into a tale of decline and fall. This tendency has recently appeared in Fredric Jameson's highly influential popularization of Ernst Mandel's analysis of the three stages of capitalism: market capitalism ("machine production of steam-driven motors since 1848"); monopoly capitalism ("machine production of electric and combustion motors since the 90s of the 19th century"); and postindustrial or multinational capitalism ("machine production of electronic and nuclear-powered apparatuses since the 40s of the 20th century"). See Fredric Jameson, *Postmodernism, or, The Cultural Logic of Late Capitalism* (Durham, N.C.: Duke University Press, 1991), 35. See also Fredric Jameson, "Postmodernism and Consumer Society," in *The Anti-Aesthetic: Essays on Postmodern Culture*, ed. Hal Foster (Port Townsend, Wash.: Bay Press, 1983).

2. Quoted in Camilla Gray, *The Russian Experiment in Art: 1863–1922* (New Haven, Conn.: Yale University Press, 1987), 1.

3. Gertrude Stein, *Picasso* (New York: Scribner's, 1939), 11.

4. Stephen Kern, *Culture of Time and Space* (Cambridge: Harvard University Press, 1989), 259–60.

5. Quoted in Robert Hughes, *The Shock of the New* (New York: Knopf, 1967), 43.

6. Quoted in ibid., 191.

7. Quoted in Robert Fishman, *Urban Utopias* (New York: Basic Books, 1977), 240. It would be a mistake to limit the relation between modernism and fascism to the Italian scene. In Germany, National Socialism cultivated a sophisticated aestheticization of politics. Indeed, the political power of the interplay between art and media on a global scale was first realized in Germany. These developments can be traced to the notion of the *Gesamtkunstwerk*, which emerged in Weimar and was developed at the Bauhaus. For further consideration of this point, see Mark C. Taylor, "Theoaesthetics," in *Disfiguring: Art, Architecture, Religion* (Chicago: University of Chicago Press, 1992), and "The Virtual Kingdom," in *About Religion: Economies of Faith in Virtual Culture* (Chicago: University of Chicago Press, 1999).

8. Andy Warhol, *The Philosophy of Andy Warhol* (New York: Harcourt Brace, 1975), 92.

9. Jonathan Crary, *Art after Modernism: Rethinking Representation*, ed. Brian Wallis (New York: New Museum of Contemporary Art, 1984), 294.

10. Ibid., 292.

11. Ibid., 294.

12. Eric Davis, *Techgnosis: Myth, Magic, and Mysticism in the Age of Information* (New York: Harmony Books, 1998). See also Mark Dery, *Escape Velocity: Cyberculture and the End of the Century* (New York: Grove Press, 1996).

13. N. Katherine Hayles, *How We Became Posthuman: Virtual Bodies in Cybernetics, Literature, and Informatics* (Chicago: University of Chicago Press, 1999), 48, 49.

14. Robert Bellah, "Freedom, Coercion, Authority," *Academe* (January/February 1999): 18.

15. Ibid., 19.

New Media and the Natural World: The Dialectics of Desire

Robert Markley
West Virginia University

A N earlier version of this essay was solicited for a collection that languished for several years. After months of struggling to revise this antediluvian (ca. 1994) paper, I scrapped most of it in 1999 to confront the problems posed by the last half decade of technological and cultural change. As the twenty-first century arrived, I again had to revise this piece in order to keep pace with a range of technological transformations which have rendered most of the mainstream pronouncements about cyberculture from the 1990s unbearably quaint if not downright antiquarian. My original intention had been to demystify the evangelical rhetoric of the early 1990s which, you no doubt recall, heralded the Internet as a revolutionary leap in human consciousness and community. Now, on the eve of another great leap forward in information technologies—from text-oriented protocols and static multimedia to dynamic video and immersive interfaces—the clamor for hypertext, the Web, and virtual reality seems the echo of a bygone era, a casualty of a digital marketplace saturated by the language, values, and assumptions of late capitalism.[1] To underscore how the rhetorical scene has shifted since the early 1990s, I offer this anecdote: at the annual Apple Developers Conference in May 1998, a session on multimedia authoring tools featured a keynote speaker (a prominent software engineer), who derided hypertext and HTML as "sandbox technologies" and prompted the audience to laugh uproariously by reading passages by humanists and social scientists which declared that the Internet recasts democratically the relationships among identity, technology, and socioeconomic power. This essay, in one respect, is an effort to explain both that laughter and the ironies that underlie it, to explore critically our faith in the planned obsolescence that drives the new media revolution.

In their efforts to ride the crest of the digital wave, producers, critics, and consumers of new media confront a fundamental paradox: the more rapidly information systems change, the more prone we seem to revisit the oppositions that have structured our understanding of technology for (at least) several centuries: technological innovation versus "inherent" human nature, expertise versus craft-labor, investment capital versus wages, and managerial control versus democratic dissemination, to name only a few.[2] In practice, we face a standoff. While technophiles may have muted their evangelism for the transformations to be wrought by new media, they have not developed a coherent vocabulary—theoretical or philosophical—to describe the technological or social effects that they are in the business of engineering; neither, however, have those cultural critics who remain skeptical of what Donna Haraway termed in the 1980s the "infomatics of domination."[3] In fact, despite their differences, both defenders and critics of cyberculture share a common heritage of dialectical approaches to technology and nature, a heritage that seems to dead-end in various myths of human-machine synthesis. My purpose in this essay, in part, is to argue that our fascination with technoculture is itself part of a dialectic—an ideology of modernity—that discourages our thinking seriously about the relationships between new media and the socioeconomic conditions under which they are produced, upgraded, and reproduced.[4] At best, this fascination with net culture fosters ways to rethink the relationships among rapidly proliferating generations of new media; at worst, it reinforces the oppositional structures of thought that divorce nature from culture and separate "essential" human qualities (the mind and spirit) from the technological fixes that characterize our implication in complex socionatural ecologies.

To work through the dialectical logics described by the oppositions human/technology and nature/culture, I begin by redefining our conceptions of how media develop, then go on to consider the implications of such a redefinition. In the first section, drawing on the work of David Bolter and Richard Grusin, I argue that all media *remediate* (to use their term) a complex archaeology of assumptions, values, and representational strategies in efforts to approximate ever more closely an unmediated reality.[5] As Marshall McLuhan reminds us, old media do not disappear: they are subsumed archaeologically, coexist with, or compete against new and emerging media forms, becoming the very "content" that new media represent.[6] The logic of remediation has profound consequences for our understanding of the relationship between new media and the nonbitmapped world; by historicizing technologies of representation, it focuses on the gap between our fascination with new media and our desire for unmediated experience. In the second section, I look critically at the dialectical logic which reinscribes variations of the posthuman—the concept, practices, and metaphors of technotranscendence that emerge each time we try to reconcile the oppositions of human/technology and nature/culture. To think beyond the limitations of dialectical synthesis, I draw on the work of A.-J. Greimas to explore the neglected "fourth term" that insistently undermines binary systems of logic and analysis. Greimas's semiotic square describes a dynamic that reframes the structures of meaning by which we make sense of the world, and, in the process, calls into question our default tendencies to celebrate our technologically sophisticated present at the expense of our predigital and devalued past. It allows us to explore what is repressed or excluded by visions of a posthuman future, notably humankind's irrevocable implication in complex, nonanthropocentric ecological systems.

The precarious balancing acts of organisms and environments that we call "ecology," however, are themselves embedded in complex matrixes that generate ongoing reconceptions of the initial terms of the dialectic: human, technology, and posthuman. In the third section, then, I explore the transformations of the nature/culture opposition to examine how technoculture downplays or blackboxes the means by which we calculate and extract value from the material world, rechanneling our desires for continually expanding resources. This faith in *simulation*—the desire to program into existence an abundance of resources which we can then exploit—needs to be seen within the context of age-old attempts by human cultures to deal with crises of intensification, the ongoing process of developing new means to secure ever more remote, expensive, and labor-intensive resources.[7] As this brief description may suggest, I want both to challenge the belief that we can program our way out of social, economic, and environmental problems and to contest conceptions of culture, nature, and technology that reduce humankind and the environment to malleable forms awaiting (or dreading) the soul-changing transformations wrought by new media.

I

Bolter and Grusin argue convincingly that all media cannibalize and subsume previous technologies of representation in their efforts to reproduce a true or authentic experience. The desire for verisimilitude, they maintain, characterizes an ongoing quest to transcend mediation. Whether we multiply such technologies to achieve unprecedented standards of visual, auditory, or sensory quality or design systems that try to remain as inconspicuous as possible, the ultimate goal is the same: "hypermedia and transparent media," according to Bolter and Grusin, "are opposite manifestations of the same desire: the desire to get past the limits of representation and to achieve the real."[8] This desire to transcend representation, however, is governed by the irony that our perceptions of "reality" are always shaped by the expectations we have derived from previous generations of media. Digital television, for example, provides images that seem more "true-to-life" only because we are comparing them to the picture quality provided by last year's models. We can describe "revolutionary" improvements in verisimilitude *only* by judging new media against the perceived inadequacy of those media that they are designed to replace. "Each new medium," Bolter and Grusin assert, "is justified because it fills a lack or repairs a fault in its predecessor, because it fulfills the unkept promise of an older medium." "In each case that inadequacy is represented as a lack of immediacy."[9] Our expectations about what constitutes immediacy, then, are structured by our perceptions of an always mediated reality—a reality, in other words, that is always in the process of becoming obsolete.

Because "reality" becomes an *effect* of technologies that constantly redefine immediacy, our desire for verisimilitude leads us to develop high-end systems that promise ever closer approximations of unmediated experience. Reality, in this respect, signifies our paradoxical desire to employ sophisticated media to transcend the belatedness of representation: it marks both the ideal of absolute fidelity or identity (is it live or is it Memorex?) and a standard that allows us to evaluate the verisimilitude of successive or competing representational technologies.[10] The implication of Bolter and Grusin's argument, then, is that "reality" and "immediacy" must be rendered as technoreality and technoimmediacy; as Michelle Kendrick puts it, "technology has always been an affective agent" in the "dialogical relationship" between technosystems and human beings.[11] The arguments of Bolter, Grusin, and Kendrick, then, imply that the purpose of new media is precisely to redefine the real and to render our desire for immediacy omnivorous and insatiable in order to justify continuing efforts to transcend the limits of representation. The fact that "reality" remains an unrealizable ideal when it is translated into the goal of representation is *not* to relativize the existence of the material world but to emphasize that our perceptions of the countless synergistic and competitive systems that we call the physical universe are constrained by the limitations of our cognitive hardware and software and by the systems of representation we construct.[12] Because the real is always a function of technological mediation, no standard of verisimilitude (number of pixels per screen, storage capacity, RAM, or processing speed) or technological breakthrough can ever render media completely transparent. Consequently, the desire for the real depends on an ever expanding use of financial resources, raw materials, and intellectual and manual labor to approach asymptotically the ideal of unmediated experience.

As new media shatter old limits on storage and speed, they also reinvent standards of verisimilitude, even experience itself. The logic of remediation drives home McLuhan's point: the contents of new media forms are the old media—that is, our perceptions of "real life"—that they have supplanted. If you're skeptical, watch what happens over the next decade. Universities currently are investing in the fiber-optic networks of Internet II, increasing data rates from 1.2 to 25 mb/s and beyond. This exponential increase will not necessarily mean an unlimited expansion of our ability to communicate vast amounts of information or even increased access to existing hypermedia technologies, as some developers intimate; rather, Internet II heralds a massive remediation of previous "passive" media—television, video, film—as "new" interactive or immersive technologies. If the products now being released are any indication, the new software designed to enable and exploit these much higher data rates will concentrate on improving video quality. In this regard, the transition from MPEG 2, to MPEG 4, to uncompressed video ratchets upwards conceptions of video verisimilitude; as the acronym M-P-E-G (Motion Picture Expert Group) suggests, the industry standard for video compression reproduces the genetic codes of the entertainment industry. A sign of the (coming) times—the data rate for uncompressed video matches the original standards, since exceeded, for the fiber-optic backbone of Internet II.

While Internet II is being touted for its potential to transmit huge amounts of scientific and technical data, the advent of high-speed cable systems allows users to access Web sites and download texts, static images, and compressed (artifact-laden) video much more quickly than they could in the era of dialup modems. As Bolter and Grusin's account of remediation suggests, it seems likely that the

World Wide Web increasingly will be dominated by high-end, expensive-to-produce, dynamic content. Current default Web sites—static text and layout that include brief multimedia clips—will become relics, replaced by locations that offer high quality video, sophisticated Web objects, and three-dimensional virtual environments. As today's graphic designers give way to would-be directors, what we now experience as state-of-the-art Web sites will be replaced by (let's face it) commercials, virtual tours of the insides of appliances, and videos of your nieces and nephews opening their Christmas presents. These technological developments do not herald a technomillennium. They will benefit those who have the knowledge, capital, time, and expertise to design and exploit them, and, in turn, they will drive the remediation of previous technologies as the engine for developing yet newer media.

Bolter and Grusin's account of remediation also provides a way to analyze the limitations of those speculative narratives that credit new media with bringing about revolutionary breakthroughs in human consciousness. In the 1980s the idealization of cyberspace as a realm of limitless possibility and transsubjective creativity led many observers to predict a dawning age of unprecedented creative freedom and self-actualization.[13] The new media revolution may seem more rhetorically pragmatic, but it too recycles a suspect faith in progressivist narratives of technology, history, and human development. In a relatively recent special issue of *Proceedings of the IEEE* devoted to multimedia, for example, Ryohei Nakatsu describes a complex research program to develop a new generation of virtual media—cyberspatial movies in which users can interact dynamically with characters and environments.[14] The emotion-recognition software that Nakatsu and others are developing begins by bitmapping and then simulating a semiotics of muscle movements and facial expressions which serve as templates for affective responses to users' input. To envision what he calls this "realm of new experience," however, Nakatsu places this sophisticated software within the context of traditional narratives that allow users of this system to "experienc[e] the world of the interactive story as heroes."[15] Regardless of how cutting edge the emotion-recognition software he describes may be, his projected system as a whole remediates conventional, often explicitly masculinist, conceptions of heroism from older media: television, science-fiction movies, sword-and-sorcery novels, comic books, computer games, and even medieval romance. In this respect, the constitutive metaphor of the immersive system he describes recalls the technoimaginary of the holodeck on *Star Trek: The Next Generation.*[16] What results is a hackneyed notion of heroism—complex psychologies are simulated, and simplified, as algorithmic responses to fixed, programmable stimuli forever disengaged from cultural, national, political, and ethnic contexts. To act as a hero in a future interactive movie, one will have to respond appropriately within the parameters that the software is designed to accommodate. Confronted by a holodeck adversary, the stripped-down self-as-hero presumably will have to fight rather than launch into an analysis of Aristotle's concept of a just society. Rather than the embodied "technesis" envisioned by latter-day romantics such as Mark Hansen, we are left with a new encoding of male hysteria: the idealization of "technology beyond writing" as the herald of yet another "revolutionary" effort to augment an always inadequate physical existence.[17]

The basis for Nakatsu's architecture, in an important sense, is his faith in the ability of technology to approach the real by transcending the perceived limitations of previous generations of media: movies are more "real" than books because they provide a "fuller" sensory experience; virtual environments are more "real" than passive media because they engage senses other than sight and hearing. Sensory experience itself, however, is judged by standards of verisimilitude derived from an engineering definition of immersive experience: more stimuli, greater processing speed, better sound quality, higher visual resolution. Given this reflexive definition of technological progress, future multimedia developments will always herald themselves as "improvements" because they will access more of what Novak terms the "full human sensorium."[18] This is not to criticize the technical viability of emotion-recognition software or the integrative technologies envisioned by Nakatsu's project. Instead, I want to emphasize that without a sophisticated understanding of remediation, we remain liable to blackbox the complex relationships between all media and their cultural and socioeconomic contexts and consequently reinscribe the naive formulations of "technesis" offered by Hansen and others. And without investigating these contexts, assumptions about gender, identity, technology, and ecology remain set to the simple default modes that we call ideology.

One of the virtues of Bolter and Grusin's analysis is that it gestures beyond a closed circuit of representation toward an ongoing critique of the oppositions—form versus matter, mind versus body, male versus female, and nature versus culture, among others—which have characterized Western philosophy since Plato.[19] Quite simply, our dialectical desires to transcend representation and return to a pristine, unmediated existence are deeply embedded in structures of thought and perception that lie at the heart of Western conceptions of self-actualization and the good life. In this respect, hypermedia and transparent media both extend and embody what Don Ihde calls the "doubled desire" that characterizes human relationships to technology.[20] On the one hand, ever more complex technologies improve our quality of life by increasing the production and distribution of goods and services; on the other, the more these technologies multiply, the more they try to render themselves transparent: their purpose is to recreate an enhanced version of a "natural," even prelapsarian, existence in which humankind's managerial classes live in a world of ease and abundance. The logic of this double desire, like that of remediation, depends on a view of technological progress that downplays the relationships between such improvements and the labor, resources, and capital they require.

Desires invariably require trade-offs: the more time, energy, and resources are invested in extracting raw materials and manufacturing finished goods, the more pressure is brought to bear on human and material resources to fuel these efforts to improve or maintain living standards. This is the spiraling logic of intensification that, Harris argues, is "always counterproductive": even "successful cultures [which] invent new and more efficient means of production" sooner or later exhaust readily accessible resources and again must face the social, economic, political, and ecological consequences of their actions.[21] Significantly, such resources are never fixed stockpiles but the available best guesses about demographic trends, patterns of consumption, strategies of allocation, price levels, fuel and transportation costs, and food supplies. Bolter and Grusin's description of remediation, then, allows us to extend Ihde's insight: because technologies of representation are always in the process of reconfiguring our sense of the real, they invariably alter perceptions of what human communities "need" to exist. Our understanding of how we relate to our environments are mediated by *contested* representations of the available stocks of resources which we call "nature."

II

Taken together, the works cited in my first twenty-odd footnotes trace minihistories of recent efforts to come to terms with two versions of the dialectics of desire. They might be represented as follows:

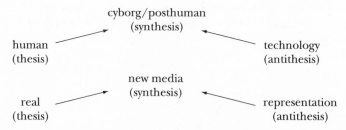

The problem is that no matter how long one ponders or reconfigures these schematics, the terms themselves seem unstable and their relationships uncertain: how can one separate "essential" human qualities from the tool-using ability we call technology? isn't the category of the human already split by the mind-body problem? and so on. The double desires that destabilize the relationships between technology and nature and between reality and representation call into question the dialectical logic which underlies the persistence of these binaries.[22] The very instability of these oppositions and others like them suggests, according to Greimas, "the existence, beyond the realm of the binary, of a more complex elemental structure of signification."[23] This "complex elemental structure" is Greimas's attempt to map the dynamic relationships, and the variety of effects, that arise from *within* dialectical logic. I invoke Greimas, then, not to impose an ironclad theoretical model on the problems of technology and representation but to explore the structures of meaning that precede and inform *both* our fascination with hybrids—such as the cyborg—that sell themselves as "solutions" to postmodern crises of representation *and* our societal reluctance to face environmental problems larger than our recycling bins. Historically and theoretically, technologies of representation are responses to, and engines of, ongoing socioeconomic, demographic, and ecological transformations. And, as anyone versed in historical ecology recognizes, the history of humankind during and after periods of rapid population growth, expansion, and technological innovation are marked by contentious efforts to manage the often devastating consequences of intensification.[24] The dialectics that underlie crises of representation and environmental degradation are isotropic, and the semiotic square provides a heuristic means to analyze the tensions that define them.

   The example that Ronald Schleifer gives—the signifying system of colors defined by the poles of "black" and "white"—demonstrates how the process of uncovering an "elemental" structure of meaning can help us understand the complex dynamics that animate oppositional logic.

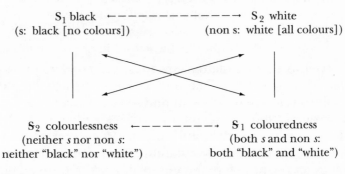

S: "particular colours"

S₁ black ←————————————→ S₂ white
(s: black [no colours])   (non s: white [all colours])

S₂ colourlessness ←————————→ S₁ colouredness
(neither s nor non s: (both s and non s:
neither "black" nor "white")   both "black" and "white")

S: "colours"

   The opposition of black and white—what Greimas calls the "contrary relationship"—engenders and is engendered by a more complex and elemental structure of signification. In the lower right corner of the square lies the "positive term," the synthesis of the Hegelian dialectic. This gesture toward transcendence, however, invariably produces the "contradictory" relationship that leads to the "negative complex term" in the lower left corner: the negation, neither black nor white, that, in Schleifer's words, "contaminates the purity of the logical syntax."[25] This term destabilizes the fixed categories "black" and "white" and signifies a radical disruption of the terms of the initial opposition. Nancy Armstrong succinctly describes this process: "Once any unit of meaning is conceived, we automatically conceive of the absence of that meaning, as well as an opposing system of meaning that correspondingly implies its own absence."[26] Simply put, the lower left position reveals that its "positive" contradiction—the Hegelian synthesis—expresses a *desire* for transcendence rather than an ultimate reconciliation of the oppositions it seeks to resolve. This complex negative term paradoxically precedes and structures the signifying systems which try to repress or expel it—what Julia Kristeva calls "the fourth term" of Hegel's dialectic, "that which remains outside logic . . . heterogeneous to logic even while producing it through a movement of separation or rejection."[27] Put simply, the lower left position reveals the unrepresentable "problem" that the initial opposition is intended to banish.

The semiotic square, as Latour recognizes, describes the paradoxes that structure what he calls the "double constitution of modernity"—the seemingly fundamental distinctions between the practices of purification (the desire to separate nature from culture) on the one hand and mediation (the efforts to synthesize this opposition) on the other.[28] The dialectic relationship between an "essential" human nature and technology represses a more "elemental" set of relationships that undergird the double desires of modernity.

Modernity

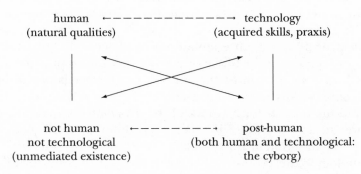

human ←–––––––––––→ technology
(natural qualities)   (acquired skills, praxis)

not human   ←–––––––→   post-human
not technological   (both human and technological:
(unmediated existence)   the cyborg)

Latour: *We Have Never Been Modern*

Mapping the opposition human/technology allows us to explore the implications of the cyborg and the posthuman more generally. The cyborg is not a radical irruption of technology into a previously coherent human psyche or body but the necessary "solution" to what is perceived as an *originary* alienation. This alienation is effectively double: in order for technotranscendence in the lower right position to function as "both/and," the human must be divided *from* an external nature and *within* itself. In this respect, the posthuman is designed to augment or repair these gaps or inconsistencies within the category of the human because, ironically, it is only through this dialectical synthesis that the originary double alienation of the human becomes evident.[29] To take only one example, consider Nakatsu's description of "metamorphoses technology."[30] To justify the expense of such metamorphoses, this immersive regime must define retroactively our experience of an unmediated, pretechnological or undertechnologized world as doubly deficient: in real life we can master neither our environment (people respond to us in unprogrammed ways) nor our selves (we can become conventional "heroes" only in cyber narratives). But the heroism that is intended to repair deficiencies in our unmediated experience is itself a bare-bones fiction: the agency which such "metamorphoses technologies" conjure into being exists only as a a wish fulfillment projected from, as well as into, remediated virtual realms.

As a heuristic, then, the semiotic square allows us to explore those characteristics that the oppositions of mind and body, nature and culture, mystify or obscure: the significance of labor and the appropriation of labor (lower right) for our conception of human nature and the co-implication of the human in complex ecological systems (lower left). The lower right position marks the posthuman as a category of strategies and practices that, by enhancing our mental and physical abilities, "liberate" us from subsistence existence. But in the reciprocal movement engendered by dialectical logic, this synthesis also reinforces the initial opposition to which it provides a "solution"—the separation of the spirit from the needs of the body. The cyborg, in this respect, marks the conservation of human intention and the human form (think of the *Terminator* movies) within a technological imperative: increase your ability to work more efficiently. As an emblem for a dialectical synthesis of "human" and "technology," the cyborg has become the poster child for Silicon Valley's perpetual revolution—the transcendent "solution" of a postindustrial economy to the tensions between labor and management.[31]

But this is only half the story. As a changeling spirited away from its military-industrial cradle by science fiction novelists, cultural critics, technofeminists, and, of course, Hollywood producers, the cyborg, as Haraway maintains, can function as an "ironic political myth" precisely by revealing and contesting the dynamic relationships that question the coherence of the human. Its antithesis in the lower left position is the double negation of the human (the natural) and technology (the artificial)—ecological systems that decenter human beings as the Baconian masters of nature. This nonanthropocentric "fourth term" challenges the very logic which summons the posthuman into existence.

As Schleifer argues, the complex negative term, by exploding the logic of the initial dialectic, invariably generates another square, another set of complex, contrary, and contradictory relationships.[32] Ecological systems may disrupt the logic that opposes the natural to the artificial, but once we try to represent them (whether in natural histories or by quantitative descriptions), we find ourselves confronting another level of dialectical logic which both complicates and enriches our understanding of the initial opposition:

Coimplication: human/nature

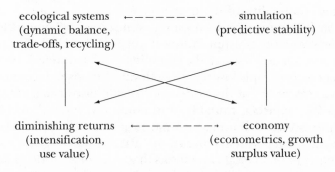

ecological systems ⟵ — — — — — — → simulation
(dynamic balance,   (predictive stability)
trade-offs, recycling)

diminishing returns ⟵ — — — — — — → economy
(intensification,   (econometrics, growth
use value)   surplus value)

Coimplication: Historical Ecology

The opposition of ecological systems and simulation is presented, in part, as a solution to the problem of representation. In contrast to Lacanian or Derridean accounts of representation predicated on lack (the absence of the signified) or deferral (the gap between sign and signified), simulation, as Steve Shaviro argues, "precedes its object: it doesn't imitate or stand in for a given thing, but provides a program for generating it. The simulacrum is the birth of the thing, rather than its death."[33] As the "contrary" to ecological systems and as the alternative to the real, simulation offers a mathematically based system that provides an ordering principle which precedes, predicts, and certifies the authenticity of the material world we inhabit. The defining characteristic of simulation, in this regard, is not the three-prong plug at the end of your computer, but the mathematics of algorithmic progression, the codes that render mathematical modeling and analysis coherent. Reality becomes the execution of a program: "We trust and believe that the world is real," Shaviro suggests, "precisely because we know it to be a simulation."[34] If such a statement sounds as though we've fallen through the postmodernist looking glass, we need to keep in mind that the guru of simulation is Leibniz. In marked contrast to Newton and the tradition of British voluntarism, Leibniz defines the Deity in terms of omniscience rather than omnipotence: we believe in God, he maintains, as an omniscient programmer who has already ensured that all possible actions have been incorporated into the program that we call existence or history before creation.[35] As Foucault recognizes, Leibniz is central to an Enlightenment mind-set, "mathesis," which extends beyond the operations of mathematics to include a science—and ideology—of order.[36] Rather than a theoretical buzz word, simulation is the centuries-old mark of faith in the ultimate computability, and rationality, of the universe.

If we lack Leibniz's faith, however, simulation becomes the programmer's version of what Haraway terms "the God-trick"—the identification of one's situated perceptions with a timeless, transcendent objectivity.[37] A philosophy that elevates the execution of a program to a guarantee of the coherence of the universe, she implies, devalues an imperfect, mutable, and feminized natural world. The lower right position illuminates what is at stake in the standoff between always intentional simulation (after all, a program needs a programmer) and ecological systems that operate outside of and beyond human intention. The "both/and" of ecology and simulation is the realm of economy, figured both as the means to manage complex ecological systems as resources and as the mathematico-technical knowledge to *produce* goods and services indefinitely into the future. Its antithesis in the lower left position is a natural world *already* in the process of being used up, characterized by constraints, scarcity, and the complex negotiations for resources, power, and profit that define political economy.[38]

What unites the terms—simulation and economy—on the right hand side of this square, then, is a version of Foucault's "mathesis," a belief in systematicity itself.[39] As the "contrary" to ecological systems, simulation envisions an alternative "reality" to guarantee systemic consistency. By controlling initial parameters, by defining limits, variables, and rates of change, simulation offers the prospect of converting unruly systems into computable entities. The logic of simulation allows us to bring ecological systems within the mathematical realm of econometric models. The right hand terms of this square, then, rely on the predictive capacity of mathematical functions carried out, in effect, to infinity. The dialectical "solution" provided by an economy of endless growth is made possible by a faith in a mathematicized, that is, ordered, universe. The lower right position, in this regard, marks the regime of a surplus value that can be extended indefinitely. As an idealization, the economy of "both/and"—ecosystems *and* simulation, the "external" world *and* human intention—offers an anti-ecology of infinite exploitation, an economics (characteristic of but not limited to late-twentieth-century capitalism) which assumes that humankind can extract or manufacture value endlessly.

The basis for this faith is buried in our assumptions about the nature of mathematics (let the groans begin). But it seems worth uncovering a few of these assumptions because they indicate why most cultures, to the bitter end, have believed that they could produce their way out of economic and ecological crises. The right hand terms of the square—simulation and growth—share the neoplatonic assumptions of what Brian Rotman calls the mathematics of "infinitarianism."[40] For Rotman, the belief that mathematical objects and truths are inscribed in the universe (awaiting our technologically enhanced ability to discover them) produces a decontextualized, disembodied mathematics predicated on "the fantasy of the absolute: absolute sameness, identity, equality, and permanence of mathematical objects for all times and all conceptual regimes and all historical materializations."[41] By extrapolating present conditions or rates of growth and production into the future, an infinitarian mathematics creates an objective, decontextualized virtual space in which we can predict the behavior of selected variables with great accuracy. The problem is that the process of selecting which variables to compute is always restrictive, always dependent on current technologies of representation and on our always limited comprehension of conditions that do not lend themselves to calculation. The "fantasy of the absolute" is, in an important sense, an expression of our desire to be able to discount what we do not understand. If we bring some components of ecological systems under a regime of calculation by labeling them as resources—as key variables in a simulated future— then other aspects of these systems must be considered as intangible or mathematically insignificant. Mathematics itself is subject to the logic of remediation: we can compute only those objects and dynamics that can be rendered within a generalized, transhistorical semiotics. The virtual space of absolute consistency, in this regard, identifies infinitarian assumptions with the desire for infinite productivity, then projects those assumptions, and that desire, onto the natural world. Put simply, an infinitarian mathematics produces inexhaustible resources to count. Seen from the perspective of an economy committed to indefinite growth, nature becomes an embodiment of human desire—a realm of inexhaustible resources that is always exploitable.

Because it exists as the negation of diminishing returns, the positive complex term—surplus value—irrevocably has both "real" and "symbolic" dimensions: it must produce solutions to real-world problems (who is going to buy stock in Always-New Media, Inc. if we go public?) and displace its ultimate transcendence of material constraints into an imaginary realm of infinite plenitude. Traditionally, the fiction of an inexhaustible and infinitely exploitable nature is projected either into the past—the "golden age"—or onto the promise of unexplored or underexploited territories: America, Africa, the Amazon; the final frontier of outer space; and, as space exploration has proved dangerous and expensive, cyberspace. "Thus in the beginning," John Locke declares in his *Second Treatise of Government*, "all the World was *America*"—that is, the world was an unmarked and endlessly exploitable territory in which resources could be used without being used up.[42] The fiction that output can exceed input, that value can be created is the age-old dream of alchemy—the expression of boundless desires predicated on our faith in a bounteous nature and infinite productivity, a simulation that seems to have been running since the "dawn" of civilization, when the needs for fuel, food, and shelter for an expanding population led to the deforestation of Mesopotamia. We may never have been modern, but, to the extent we believe in what Paul Hawken calls the "pervasive untruth" of contemporary capitalism, the belief that "economic growth can be extended indefinitely into the future," we have always been virtual.[43]

As the regime of surplus value, the lower right position offers a "solution" to the problem of *desire* by making desire endlessly productive.[44] In the early 1990s, some commentators intimated that cyberspace would transcend traditional economics. Michael Benedikt, for example, predicted that "the economic principles of material production and distribution in their classically understood forms—principles of property, wealth, markets, capital, and labor—are no longer sufficient to describe or guide the dynamics of our modern, complex, 'information' society."[45] Assertions such as Benedikt's depend on the assumption that electronic technologies can produce resources—namely, information—which then can be exploited for profit in the place of material goods, practices, and embodied skills. Such technologies of enablement, the theory goes, allow postindustrial society to bypass ecological, political, and economic constraints on labor and resources and manufacture value indefinitely. This schematic of Electronic Entrepreneurship 101, however, does not explain why efforts to find venture capital for high-tech start-ups are successful less than one percent of the time.[46] Nor does it describe the actual experience of the competitive world of high-tech financing in which the norms seem to be cutthroat competition, rapidly obsolescent software, and market predictions gone south (the seven-billion dollar a year industry in home virtual reality gear that was supposed to exist by 2000).

Rather than redefining the basic principles of economics, new media intensify our culture's faith in "the unlimited increase of objects" to buy, sell, and desire by "present[ing] the illusion of choice on a massive scale."[47] The Internet markets desire itself, reinforcing a familiar capitalist dream of ongoing commodification in which "desire multiplies to match the ceaseless multiplication of things."[48] New media exploit not simply focused desires for new products or technological breakthroughs but our fascination with new modes of distribution and with the volatility of a market in which commodities are always being discounted, revalued, pirated, bought, sold, rendered obsolescent, and discarded. As the antitrust suit against Microsoft's efforts to make its Internet browser an industry standard demonstrated, traffic on the World Wide Web *is* the commodity. The driving force in the Web's transition from static HTML sites to dynamic and transactional content, according to Larry Slotnick, Apple's Vice President for Internet and Enterprise Products, is "automated self-service" for consumers.[49] Collectively, we imagine these transactions taking place in a dematerialized, virtual space with its own systemic logic, not to mention its own security system to guard our credit card numbers. But traditional economic principles have not radically altered, and the plunge of the NASDAQ in 2001–02 has put an end to recurring fantasies of capitalism transcending "business cycles"—the fiction that numbers and charts correspond to human behavior and an ecologically independent "economy." Cyber transactions reinforce the belief in abstract economic "growth" and the anti-ecological premise that the production and consumption of goods and services will continue indefinitely. As the velocity of money increases, the desire *to* desire projects these turbocharged principles back into our perceptions of "real life." In the world of e-commerce, the self becomes a function of desire, *homo economicus* with high-speed cable access.

III

The idealized visions of endless growth and inexhaustible resources also produce, however, contradictory perspectives on the very technologies of representation that bring them into being. Mathematics functions as both a metaphysic and a method, a transcendental standard of value and an instrumental technology. In the lower left position, operating as such a technology, mathematics describes a regime of scarcity. Calculation and extrapolation start precisely when resources become resources, when they become *worth* counting. In this respect, mathematics marks another mythic origin—the fall from abundance that renders its technologies co-extensive with the need to represent, count, measure, estimate future stores and requirements, calculate rates of use and expenditure, allocate, hoard, and exchange. The mathematics of intensification, however, does not entail a retreat to a mindless Malthusianism, which naively inverts the multiplying commodities of infinite production into endless shortages and arrested technological development. The "contradiction" of diminishing returns instead explodes the idealized notions framed by the mathematics of infinite productivity. Intensification recasts mathematics as an embodied semiotics, computations that take place in historical time and represent historical strategies and solutions, not absolute truths.[50]

The contradictory functions mapped by the lower left and right positions represent the impossibility of either "pure" mediation or "pure" separation, the "pure" abstraction of econometrics or "pure" unrepresentability of a transcendent nature. The very acts of calculating logging expenditures, number of board feet per tree, wages, profits, and cost-benefit analyses of reforestation create a virtual space. This space—precisely because it is "featureless," abstract, and general—encourages the development of newer, more sensitive, mathematical technologies to estimate and calculate resources that are not restricted to historically bound measurements. As both a transcendental guarantee and an instrumental technology, mathematics ensures that we do not have to determine the costs of cutting down each tree in a two hundred acre stand; we can average costs, calculate profits, and estimate future expenditures on reforestation. But if this abstract space of cost-benefit analyses dematerializes a complex ecology, the "real" trees resist absolute determinations of their current or future value. Prices fluctuate, laws intervene, and the trees are affected in unpredictable ways by the thousand natural shocks that plants are heir to. The virtual and the real, in short, are always dialectically bound.

The dynamic tensions that generate and are generated by productivity and intensification allow us to see why the lower left position cannot be conceived simply as a refuge for well-intentioned desires to live in "harmony" with a natural world or efforts to redirect our desires away from conspicuous consumption to "intangibles: experience, entertainment, knowledge, wisdom."[51] Our culturally embedded conceptions of harmony with nature imagine a pristine use value, a relationship with the natural world that provides us with sources for the twenty-two amino acids we need to live as well as clothing, shelter, transportation, companionship, and entertainment that are always sufficient, never excessive. But "harmony" and "sufficiency" are differential rather than absolute values; we cannot imagine or evaluate them without having experienced imbalances, desires, excesses, and shortages. Use value is a forever unexperienced state of equilibrium, a fantasy of a balance between physical needs and psychic and ideological desire. Even as it explodes the logic of the endless generation of surplus value, then, use value has no positive, nonrelational meaning. Simply put, use is always use *for*—expenditures of our time, labor, and energy that bring resources within complex semiotic systems of exchange and representation. Our perception of the natural world is always and already structured by the uses to which it will be put—whether deforestation or ecotourism. As the sign for the dynamic interaction of humankind and the environment, as a representation of Richard Lewontin's recognition that all species are continually degrading or destroying the very conditions that allow them to exist, use value implies that there can be no final, calculable account of the extraction of value from "nature" because there is no point *outside* of these ecologies from which to compute all the effects of billions of open systems—human beings—living in and off of trillions of other open systems, biological and nonbiological.[52]

As I suggested earlier, the semiotic squares I have sketched are heuristic, easily recast, disputed, or supplemented. The point, however, is not to guess what terms fit the four positions or to debate whether simulation is the precise contradiction of ecological systems; it is to recognize the relational dynamics that govern humankind's efforts as a symbol-using and resource-extracting species. Regardless of how one maps the oppositions that structure our understandings of representation and reality, the solutions promised by dreams of transcendence—the cyborg worker, the economy of infinite expansion—always prove stopgap measures. Striving for the real, even when it's rechristened "embodied technesis," returns us to the problem of remediation; focusing on representation forces

us to confront the practices which continually deform its schemes.
But no dog can chase its tail endlessly. At stake in the development
of new media is ultimately the question of whether we can interpret
critically the metanarratives—the ideology of progressivist modernity—that downplay or ignore the socioecological consequences of
our dreams of a technomillennium. The quest for an unmediated
reality allows us to displace our anxieties about environmental degradation into a dynamic of desire: the more we cluck tongues about
the state of the environment, it seems, the greater the temptation
becomes to light out for technofrontiers that grow, if such a neolo-
gism makes sense, more infinite the more we desire them.

As critics of the labor practices of Silicon Valley have noted, there
is resistance within the software and knowledge industries to a "productivity" that is characterized by postindustrial strategies of intensification—getting workers to put in eighty-hour weeks for forty hours
pay, then laying them off when the bubble deflates.[53] The new
media revolution grows ever more labor intensive. As CD-ROMs
begin to be replaced by DVD-ROMs, production and labor costs
jump severalfold, and the default model of technological innovation
as a labor-saving and cost-cutting process goes the way of the five and
a quarter inch floppy.[54] Multibillion dollar investments in high-end
multimedia applications suggest that we are witnessing the matura-
tion and flattening out of the luxury market in computers. The
phaseout of floppy drives demonstrates forcefully that our low-end
digitized past is being consigned to the scrap heap as well: software,
it turns out, has a far shorter half-life than print.[55] As productivity
rises for workers across the spectrum, the Lorenz curve, which measures the distribution of income, in the United States deforms
toward a right angle. It is still the appropriation of labor that generates "surplus" value.

The innovations benchmarked by DVD technology, uncompressed video, and Internet II may very well herald what David Porush calls a coming "cyborg illiteracy," and the end of alphabetic
consciousness may open the way for new conceptions of the semiotics of visualization.[56] But the new content search engines envisioned
by Shih-Fu Chang, Alexandros Eleftheriadis, and Robert McClintock
paradoxically require a more sophisticated theory of technoculture,
not the recycled rhetoric that characteristically trumpets the arrival
of another generation of media.[57] Such a theoretical discourse
needs to pay sustained critical attention to the excluded or repressed terms that are produced by efforts either to divorce or con-

flate "representation" and "reality": use value, labor, and intensification. In the first decade of the twenty-first century, the costs of demographic and ecological pressures are rising in ways that resist exact quantification: the planet warms, the Indonesian rain forest continues to burn, fish stocks dwindle, and fresh water supplies in many parts of the world become increasingly scarce. Remediate that.

Notes

My approach in this essay has been shaped in many ways by the four-year process of co-authoring the first scholarly DVD-ROM designed and written from scratch. See Robert Markley et al., *Red Planet: Scientific and Cultural Encounters with Mars* (Philadelphia: University of Pennsylvania Press, 2001). The history and theoretical implications of producing this 400-screen multimedia work are described in Robert Markley, Helen Burgess, and Jeanne Hamming, "The Dialogics of New Media: Video, Visualization, and Narrative in *Red Planet: Scientific and Cultural Encounters with Mars*," in *Writing Visually: Rhetoric in the Age of Multimedia,* ed. Michelle Kendrick and Mary Hocks (Cambridge: MIT Press, forthcoming).

1. See Michael Perelman, *Class Warfare in the Information Age* (New York: St. Martin's Press, 1998) and Dan Schiller, *Digital Capitalism* (Cambridge: MIT Press, 1998). For cyberenthusiasts, see the essays in *Cyberspace: First Steps,* ed. Michael Benedikt (Cambridge: MIT Press, 1991); George Landow, *Hypertext: The Convergence of Contemporary Critical Theory and Technology* (Baltimore, Md.: Johns Hopkins University Press, 1992); and Michael Joyce, *Of Two Minds: Hypertext Pedagogy and Poetics* (Ann Arbor: University of Michigan Press, 1995). For critics, see Neil Postman, *Technopoly: The Surrender of Culture to Technology* (New York: Knopf, 1992); Arthur Kroker and Michael A. Weinstein, *Data Trash: The Theory of the Virtual Class* (New York: St. Martin's Press, 1994); Stephen Talbott, *The Future Does Not Compute: Transcending the Machines in Our Midst* (Sebastapol, Calif.: O'Reilly, 1995); and Clifford Stoll, *Silicon Snake Oil: Second Thoughts on the Information Highway* (New York: Doubleday, 1995). For useful critiques of such manichean interpretations, see Richard Grusin, "What Is an Electronic Author?" in Robert Markley, ed., *Virtual Realities and Their Discontents* (Baltimore, Md.: Johns Hopkins University Press, 1996), 39–53; William J. Mitchell, *City of Bits: Space, Place, and the Infobahn* (Cambridge: MIT Press, 1996); and Hakim Bey, "The Information War," in *Virtual Futures: Cyberotics, Technology and Post-Human Pragmatism,* ed. Joan Broadhurst Dixon and Eric Cassidy (London: Routledge, 1998), 3–8.

2. On the oppositions that underlie the information age, see Donna Haraway, *Simians, Cyborgs, and Women: The Reinvention of Nature* (New York: Routledge, 1991), 149–81.

3. Ibid., 161–65.

4. The work of Martin Heidegger remains crucial for understanding the complexities of modern technology; see *The Question Concerning Technology and Other Essays,* trans. William Lovitt (New York: Harper & Row, 1977). For persuasive critiques of modernist approaches to sci-tech, see Bruno Latour, *We Have Never Been Modern,* trans. Catherine Porter (Cambridge: Harvard University Press, 1993) and Joseph

Rouse, "Philosophy of Science and the Persistent Narratives of Modernity," *Studies in History and Philosophy of Science* 22 (1991): 141–69.

5. Jay David Bolter and Richard Grusin, *Remediation: Understanding New Media* (Cambridge: MIT Press, 1999).

6. See particularly Marshall McLuhan, *Understanding Media: The Extensions of Man* (New York: New American Library, 1964) and, for a useful assessment of McLuhan's work, Arthur Kroker, "Digital Humanism: The Processed World of Marshall McLuhan," in Arthur and Marilouise Kroker, eds., *Digital Delirium* (New York: St. Martin's Press, 1997), 89–113.

7. On intensification, see Marvin Harris, *Cannibals and Kings: The Origins of Cultures* (New York: Random House, 1977) and Jack Goldstone, *Revolution and Rebellion in the Early Modern World* (Berkeley: University of California Press, 1991).

8. Jay David Bolter and Richard Grusin, "Remediation," *Configurations* 5 (1996): 313.

9. Ibid., 343, 314.

10. On representation in science, see Steve Woolgar, *Science: The Very Idea* (London: Tavistock, 1988).

11. Michelle Kendrick, "Cyberspace and the Technological Real," in Markley, ed., *Virtual Realities*, 144.

12. See Richard Lewontin, "Facts and the Factitious in the Natural Sciences," *Critical Inquiry* 18 (1991): 140–53; N. Katherine Hayles, "Constrained Constructivism: Locating Scientific Inquiry in the Theater of Representation," *New Orleans Review* 18 (1991): 76–85; and N. Katherine Hayles, "Virtual Bodies and Flickering Signifiers," *October* 66 (fall 1993): 69–91.

13. Marcos Novak, "Liquid Architecture in Cyberspace," in *Cyberspace: First Steps*, ed. Michael Benedikt (Cambridge: MIT Press, 1991), describes cyberspace as a "habitat of the imagination, a habitat for the imagination . . . the place where conscious dreaming meets subconscious dreaming, a landscape of rational magic, of mystical reason, the locus and triumph of poetry over poverty, of 'it-can-be-so' over 'it-should-be-so'" (226).

14. Ryohei Nakatsu, "Toward the Creation of a New Medium for the Multimedia Era," *Proceedings of the IEEE* 86 (1998): 825–39.

15. Ibid., 833, 832.

16. On constitutive metaphors in science, see Nancy Leys Stepan, "Race and Gender: The Role of Analogy in Science," *Isis* 77 (1986): 261–77 and James Bono, "Science, Discourse, and Literature: The Role/Rule of Metaphor in Science," in *Literature and Science: Theory and Practice*, ed. Stuart Peterfreund (Boston: Northeastern University Press, 1990), 59–90. See also Rosalind Williams, "The Political and Feminist Dimensions of Technological Determinism," in *Does Technology Drive History? The Dilemma of Technological Determinism*, ed. Merritt Roe Smith and Leo Marx (Cambridge: MIT Press, 1994), 217–35.

17. Mark Hansen, *Embodying Technesis: Technology Beyond Writing* (Ann Arbor: University of Michigan Press, 2000).

18. Novak, "Liquid Architecture," 225.

19. See Michael Heim, *The Metaphysics of Cyberspace Space* (New York: Oxford University Press, 1993); on the oppositional structures of Western thought, see Jean-Joseph Goux, *Symbolic Economies after Marx and Freud*, trans. Jennifer Curtiss Gage (Ithaca, N.Y.: Cornell University Press, 1990) and Jean-Joseph Goux, *Oedipus, Philosopher*, trans. Catherine Porter (Stanford, Calif.: Stanford University Press, 1993).

20. Don Ihde, *Technology and the Lifeworld: From Garden to Earth* (Bloomington: Indiana University Press, 1990), 75–76.

21. Harris, *Cannibals and Kings*, 5.

22. On the continuing significance of Hegelian dialectic, see Slavoj Žižek, *Tarrying with the Negative: Kant, Hegel, and the Critique of Ideology* (Durham, N.C.: Duke University Press, 1993).

23. A.-J. Greimas and Joseph Courtes, *Semiotics and Language: An Analytical Dictionary*, trans. Larry Crist and Daniel Patte et al. (Bloomington: Indiana University Press, 1982), 25. See also A.-J. Greimas, *Structural Semantics: An Attempt at a Method*, trans. Daniele McDowell, Ronald Schleifer, and Alan Velie (Lincoln: University of Nebraska Press, 1983). On the significance of Greimas's work, see Ronald Schleifer, *A.-J. Greimas and the Nature of Meaning: Linguistics, Semiotics and Discourse Theory* (London: Croom Helm, 1987) and Fredric Jameson, *The Prison House of Language* (Princeton, N.J.: Princeton University Press, 1972).

24. See particularly Alice E. Ingerson, "Tracking and Testing the Nature-Culture Divide," in *Historical Ecology: Cultural Knowledge and Changing Landscapes*, ed. Carole Crumley (Sante Fe, N.M.: School of American Research Press, 1994), 43–66; Carole Crumley, "Historical Ecology: A Multidimensional Ecological Orientation," in *Historical Ecology*, 1–16; Jared Diamond, *Guns, Germs, and Steel: The Fates of Human Societies* (New York: Norton, 1997); William Balée, "Historical Ecology: Promises and Postulates," in Balée, ed., *Advances in Historical Ecology* (New York: Columbia Univeristy Press, 1998), 13–29; and Elizabeth Graham, "Metaphors and Metamorphism: Some Thoughts on Environmental Metahistory," in *Advances in Historical Ecology*, 119–37.

25. Schleifer, *A.-J. Greimas*, 26.

26. Nancy Armstrong, "Inside Greimas's Semiotic Square: Literary Characters and Cultural Restraint," in Wendy Steiner, ed., *The Sign in Music and Literature* (Austin: University of Texas Press, 1981), 54.

27. Julia Kristeva, *Revolution in Poetic Language*, trans. Margaret Waller (New York: Columbia University Press, 1984), 112.

28. Latour, *We Have Never Been Modern*, 60–63.

29. On the posthuman, see Anne Balsomo, *Technologies of the Gendered Body: Reading Cyborg Women* (Durham, N.C.: Duke University Press, 1996); N. Katherine Hayles, *How We Became Posthuman: Virtual Bodies in Cybernetics, Literature, and Informatics* (Chicago: University of Chicago Press, 1999); and the essays in Judith Halberstam and Ira Livingston, eds., *Posthuman Bodies* (Bloomington: Indiana University Press, 1995).

30. Nakatsu, "Toward the Creation of a New Medium," 826.

31. In addition to Haraway, *Simians*, 149–81, see Chris Gray et al., *The Cyborg Handbook* (New York: Routledge, 1995).

32. Schleifer, *A.-J. Greimas*, 30–33.

33. Steve Shaviro, *Doom Patrols* (London: Serpent's Tail Press, 1996), 128.

34. Ibid.

35. On the significance of Leibniz's monadology for theorizing cyberspace, see Heim, *Metaphysics of Cyberspace* and Robert Markley, "Boundaries: Mathematics, Alienation, and the Metaphysics of Cyberspace," *Configurations* 3 (1994): 485–507; reprinted (with minor changes) in Markley, ed., *Virtual Realities*, 55–77. On Newton and Leibniz, see A. Rupert Hall, *Philosophers at War* (Cambridge: Cambridge University Press, 1980).

36. Michel Foucault, *The Order of Things: An Archaeology of the Human Sciences* (New York: Pantheon Books, 1971), 57. See also Robert Markley, "Foucault, Modernity, and the Cultural Study of Science," *Configurations* 7 (1999): 153–73.

37. Haraway, *Simians*, 193–95.

38. On the complex and inherently unstable negotiations that characterize political economy, see James A. Caporaso and David Levine, *Theories of Political Economy* (Cambridge: Cambridge University Press, 1992).

39. On the mutually constitutive forms of legitimation between mathematical representations of physical reality and of supposedly invariant economic "laws," see Philip Mirowski, *More Heat Than Light: Economics as Social Physics, Physics as Nature's Economics* (Cambridge: Cambridge University Press, 1989); Alfred W. Crosby, *The Measure of Reality: Quantification and Western Society, 1250–1600* (Cambridge: Cambridge University Press, 1997); and Theodore M. Porter, *Trust in Numbers: The Pursuit of Objectivity in Science and Public Life* (Princeton, N.J.: Princeton University Press, 1995).

40. Brian Rotman, *Ad Infinitum: The Ghost in Turing's Machine. Taking God Out of Mathematics and Putting the Body Back In* (Stanford, Calif.: Stanford University Press, 1993).

41. Ibid., 156. No person can perform, for example, the function indicated by the set of odd numbers (1, 3, 5, 7 . . .) because no one can count forever. The injunction, *list all odd numbers*, implied by this set, then, splits the mathematician into three functions: a historical figure in time; a professional who works on problems, uncovers new symmetries, and publishes her findings; and a hypothetical demon who counts, calculates *pi*, or performs endless series of functions to infinity. This projection of the neoplatonic demon, Rotman contends, has profound consequences. See also Brian Rotman, "Toward a Semiotics of Mathematics," *Semiotica* 72 (1988): 3–37.

42. John Locke, *Two Treatises of Government*, ed. Peter Laslett (Cambridge: Cambridge University Press, 1960), pt. 2, ¶s 49, 301.

43. Paul Hawken, *The Ecology of Commerce: A Declaration of Sustainability* (New York: HarperBusiness, 1993), 32–33.

44. On the production of desire, see Gilles Deleuze and Felix Guattari, *A Thousand Plateaus: Capitalism and Schizophrenia*, trans. Brian Massumi (Minneapolis: University of Minnesota Press, 1987).

45. Michael Benedikt, "Cyberspace: Some Proposals," in *Cyberspace*, 122.

46. See Richard L. Manweller, *Funding High-Tech Ventures* (Grants Pass, Ore.: Oasis Press, 1997).

47. Richard Coyne, *Designing Information Technology in the Postmodern Age: From Method to Metaphor* (Cambridge: MIT Press, 1995), 79.

48. Susan Buck-Morss, "Envisioning Capital: Political Economy on Display," *Critical Inquiry* 21 (1995): 456.

49. Untitled presentation, Worldwide Apple Developers Conference, San Jose, Calif., 13 May 1998.

50. See Lewontin, "Facts and the Factitious," 140–53.

51. James Ogilvy, "The Power to Consume," *Wired* 5 (1997): 112.

52. See Richard Levins and Richard Lewontin, *The Dialectical Biologist* (Cambridge: Harvard University Press, 1985), 133–42 and 272–85.

53. In addition to Schiller, *Digital Capitalism*, and Perelman, *Class Warfare*, see

Stanley Aronowitz and William DiFazio, *The Jobless Future: Sci-Tech and the Dogma of Work* (Minneapolis: University of Minnesota Press, 1994).

54. The total development costs for multimedia authoring on CD-ROM average $400,000 ($100,000 for content creation and acquisition; $200,000 for production; $60,000 for testing; $40,000 for overhead and miscellaneous costs). Extrapolating these costs for DVDs (the industry standard holds 4.38 GB, seven times that of a CD-ROM), one gets over $2,892,000; allowing for economies on production and testing, the costs are still in the $2,000,000 range. See Steve Cunningham and Judson Rosebush, *Electronic Publishing on CD-ROM* (Sebastapol, Calif.: O'Reilly, 1996), 259–75. Multimedia titles, as my collaborators and I found out, are far more labor intensive than sitting down and writing a book. For every hour I spend writing my book about Mars, I spent three to six hours working on my part of our collaborative multimedia project. See Robert Markley, *Dying Planet: Mars and the Anxieties of Ecology from the Canals to Terraformation* (Durham, N.C.: Duke University Press, forthcoming) and Markley et al., *Red Planet*.

55. See Amy Johns, "Mortal Media: From Data to Dust, *Wired* 6 (1998): 78.

56. David Porush, "Telepathy: Alphabetic Consciousness and the Age of Cyborg Illiteracy," in *Virtual Futures*, 45–64.

57. Shih-Fu Chang, Alexandros Eleftheriadis, and Robert McClintock, "Next Generation Content Representation, Creation, and Searching for New-Media Applications in Education," *Proceedings of the IEEE* 86 (1998): 884–904.

Doing What Comes Generatively: Three Eras of Representation

J. Yellowlees Douglas
University of Florida

> Scenes photographed in a straightforward way are presumed to have contained the people/objects depicted. Unless obviously montaged or otherwise manipulated, the photographic attraction resides in a visceral sense that the image mirrors palpable realities.
> —Fred Ritchin, *In Our Own Image*

> [Using digital technologies for manipulating photography] is like limited nuclear war. There ain't none.
> —Robert Gicka, former Director of Photography, *The National Geographic*

> From this day on, painting is dead.
> —Paul Delaroche, on first seeing daguerreotypes in 1840

SOMEHOW, between the two of them, the agency's art director and the client's director of marketing had cooked up a concept for the next brochure cover: the client's cruise ship gliding past the mouth of Venice's Grand Canal, a view framed on either side by the old familiar vista of palazzos and gondolas. So we flew to Venice and, the day before the ship sailed, had a photographic dress rehearsal on the Pont dell' Accademia, four of us toting about seventy pounds of cameras, tripods, and bags of film: we had ninety seconds to capture the perfect shot for our next brochure cover of the cruise ship perfectly centered at the end of the Grand Canal. We spent an hour nudging tripods and cameras around to ensure the right coverage, to hit all the perfect angles, then timed how long it takes to fire from all five cameras simultaneously the instant the ship glided into position. On the afternoon of the shoot, the sky cleared, the sun turning the canals a mossy agate—and we discovered a smoke-belching

58

dredger had taken up residence in the exact center of the Grand Canal, a rusted mast of steel riding on a scarred barge towering nearly the height of our cruise ship. The photographer, of course, nearly had an infarction, his tripods and bags sagging off his shoulders. "Relax," the art director told him calmly. "When we get back to London, we'll hire a Paintbox and just touch the dredger out."

At the BBC's White City studios, the director of a television drama shooting on a sound stage sends for a graphic designer who uses one of the Beeb's many Quantel Harry video-retouching systems. The drama is set in Monument Valley, but the sound stage has only a clutch of papier-mâché buttes and a floor of imitation sagebrush and local sand. Using the Harry system and a library of clips and JPEG files, the designer is able to create a series of panoramic vistas worthy of a John Ford western: buttes, canyons, desert, and apparently endless skies.

"I just wanted you to check something," the director says, drawing her onto the set and pointing to a ladder lying alongside one of the buttes, while the actors shuffle around impatiently under the lights. "Should I get one of the crew to move that? Is it going to mess everything up?" And he seems vaguely put out when she begins laughing loudly.

Behold what happens when the reproductive meets the generative. In the Paintbox suite, the art director clones the pattern of light and shadow playing on the canal surface around the dredge, then paints its shadows and hues over the barge and derrick, masking them both completely. Sitting down before her Harry system, the BBC designer begins building an entire western landscape from a library of high resolution images of Monument Valley and the Painted Desert, blending them in with the papier-mâché buttes and actors—the ladder, as she knows, is the very least of it.

The great danger of our age, some skeptics would have us believe, is that technology changes, but our habits of perception do not. The photo finishes that decide the outcome of horse races, the video replays and Abscam tapes that arrest movement and attest to guilt and help us assist fair play and nail perps alike, the dictabelt recordings that forced a president from office—all these are samples of representation, pure and simple. For the past 150 years, we had reproductive technologies that enabled us to snare an instant, preserve it, and capture reality, artifacts that could stand up in court, before congressional hearings. But now we live in an era when the *National Geographic* can digitally shuffle the pyramids of Giza to make a more picturesque group for its cover shot. Where video delivers the impossible: Natalie Cole singing a duet with her father, Elton John jamming with Count Basie, long-dead film stars cropping up in nineties ads for soft drinks. Recordings of Sibelius, Scarlatti, and Satie that represent dozens of samples, blended together in a seamless performance—with the occasional glitch patched in to give the effect of an actual, minutely flawed performance. If the very prospect of bringing together images and words to convey the nightly news has struck horror in the breast of many a Luddite from Neil Postman to Barry Sanders, we can expect the latest wave of digital technologies to induce in them either paralytic fear or apoplectic rage. Whenever, that is, the Luddites actually discover that the photograph or video they see today depicts a thing that not only never existed, but could not, in some cases, ever possibly exist. That the media for representation are changing is beyond dispute. The bag of tricks brought to the fore courtesy of digital technologies, however, is about as old as representation itself. What has lagged behind is the credulity with which we've read all forms of representation—and our ability to appreciate the potential in new digital technologies to change the ways in which we frame the world aesthetically, philosophically, epistemologically.

The Three Eras of Representation

This chaos, this uncouth and shapeless appearance, by a kind of magic, at a certain distance, assumes form and all the parts seem to drop into their proper places . . . I have often imagined that [Gainsborough's] unfinished manner contributed to that striking resemblance for which his portraits are so remarkable . . . It is presupposed that in this undetermined manner there is the general effect; enough to remind the spectator of the original; the imagination supplies the rest, and perhaps more satisfactory to himself, if not more exactly, than the artist, with all his care, could possibly have done.[1]

If we examine a work of ordinary art, by means of a powerful microscope, all trances of resemblance to nature will disappear—but the closest scrutiny of the photogenic drawing discloses a more absolute truth, a more perfect identity of aspect with the thing represented.[2]

Before we had Photoshop and Morph and Kai's Power Tools, we had Nikons and Kodak Brownies and daguerreotypes. Before we had photography, we had Sargeant and Rembrandt, Balzac and Zola. The earliest gestures toward representation were mimetic, imitative, drawings and paintings and murals that, however closely they approximated the real thing, could not quite be mistaken for it. In its early history, mimetic art strained toward verisimilitude, toward depicting something sufficiently realistically for it to stand for the things depicted. The means by which mimetic art was judged survives in the anecdote related by Pliny on the challenge between artists Parrhasios and Zeuxis. After hearing that Zeuxis had painted grapes so realistically that birds attempted to peck at the canvas, Parrhasios invited his rival to the studio to show him his own work, concealed behind a curtain. When Zeuxis attempted to raise the curtain, however, he found that Parrhasios had deceived him every bit as effectively as grapes had fooled the birds: the curtain was painted onto the panel itself.[3] "In antiquity the conquest of illusion by art was such a recent achievement that the discussion of painting and sculpture inevitably centered on imitation, *mimesis*," E. H. Gombrich writes in *Art and Illusion*: "Indeed, it may be said that the progress of art toward that goal was to the ancient world what the progress of technics is to the modern: the model of progress as such."[4]

With the debut of the daguerreotype, however, the world at last had a technology that enabled it to arrest nature, to capture reality, to preserve forever the way a person, place, thing actually looked. Reproductive art—which later included film, as well as magnetic tape and video—inextricably fused subject and object: you couldn't have a photographic or cinematic object without the original subject it depicted. Photographs became irrefutable evidence of infidelity, of atrocities committed. Photojournalists working for periodicals like *Life* sought to bring us the world via photographs; documentary films gave us things as they truly were.

Enter the computer, digitality, the image with no distinctions between original and copy, no negatives, the machine that can seamlessly move pyramids and dredging barges, and, postproduction, can surround actors on a London sound stage with all the trappings of Monument Valley. With generative art, the subject can exist, might have existed, need not have existed, cannot possibly ever have existed. Benoit Mandelbrot's fractals, which have become a staple of Hollywood's most realistic special effects, are numbers which, when plotted, produce geometric curves and forms that resemble the natural world.[5] Of course, therein lies the nefarious nature of generative technology: what it produces may look like a photograph, but it can also show us something that never happened, that couldn't possibly happen, like the pyramids on the cover of the *National Geographic*. Or, similarly, it can show us something less than the reality that existed at the instant the portrait is captured: Don Johnson *sans* the bulging shoulder holster and handgun with which he was originally photographed for the *Rolling Stone*,[6] an actor entering his sixth decade with four decades of expression lines removed, Hoffman and Cruise posing for an affable photograph together for a story on the success of *Rainman* when the actors were actually thousands of miles apart.[7]

Three ages of representation: mimetic, reproductive, generative—yet none of them entirely displaces the other. Painting and the mimetic arts, contrary to the midnineteenth-century predictions of both Delaroche and Poe (see above—while noting that this instruction itself is a hold-over from the days when manuscripts existed on scrolls), survive even as we step across the threshold of the generative age. Those who mourn the advent of digital technology and its corruption of the photograph, audio- and videotape as fail-safe windows on the world, however, are longing for a pre-Lapsarian method delivering the world intact and without intervention that never existed. Camille Silvy's *River Scene, France* (1858) may look like a particularly fortuitous meeting of society and nature. But the photographer has carefully choreographed his human subjects, posing the bourgeoisie before the private garden, the common people on the unruly, unkempt common land. The image was made from two negatives: one for the sky, presumably photographed elsewhere or on another day; one for the landscape. On developing the landscape negative, Silvy burned in the river foreground to balance the heavy, atmospheric sky. Then, to disguise the joining of the two negatives, he painted a line of cloud on the negative and printed the final result—a natural scene of casual, bucolic bliss, doctored to the absolute max.[8]

The reproductive image, it seems, has never been quite true to its subject, even from the outset. Early engravers developed a single glass plate, the *cliché-verre*, that could be scratched and painted and then used as a contact negative for producing photographic prints. Nineteenth-century photographers routinely scratched and painted negatives, dropped out backgrounds, accentuated contours, and brought up details. To keep his pioneering portrait photography studio ticking over, popular society photographer Nadar employed "six retouchers of negatives, and three artists for retouching positive prints."[9] And, at the same time the Lumière brothers were producing their cinematographic chronicles of everyday life, Georges Meliès was experimenting with cinema's first special effects in *Voyage to the Moon* by stopping the camera, and removing an actor and rolling again, making him seem to disappear in a puff of smoke.[10] In any reproductive representation, some part of the subject might exist in the world, but that doesn't ensure that the technology delivers it to us in a "pure," untouched form.

Of course, the subject itself might not have existed—at least, not in the form the photographer or artist would like us to believe. The butterfly perching on Walt Whitman's fingertip in the studio portrait of 1883, later used as the frontispiece in *Leaves of Grass*, was actually constructed of cardboard and a wire loop, later found among Whitman's effects after his death (*RE*, 196). Joe Roesnthal's famous flag-raising at Iwo Jima was also staged with a larger flag hours after the "real" flag-raising under enemy fire.[11] "Whether . . . intervention consists merely of marking, shading, and tinting in a direct print, or of stippling, painting or scratching on the negative, or of using glycerine, brush and mop on a print, faking has set in," photographer Edward Steichen wrote at the outset of his career. "In fact, every photograph is a fake from start to finish."[12]

Rosenthal's photograph at Iwo Jima—or Robert Capa's famous image of a falling soldier—cannot exist as reproductive representations without a surrounding context, one usually provided by captions. As Walter Benjamin has suggested in "A Short History of Photography," captions are the essential component of pictures.[13] Photographs, like all representations, are always part of a larger structure, one that is, as Roland Barthes insisted, "always in communication with text, caption, article."[14] The photograph can only act as our surrogate witness if an external reference, a caption or title, tells us what we're seeing. Capa's photograph was published in *Life* in 1937 with the caption "Robert Capa's camera captures a Spanish soldier at the instant he is dropped by a bullet through the head in front of Cordoba." Yet Phillip Knightley has argued that the photograph itself is so blurred and decontextualized—there's no evidence of blood, bullets, impact, or of Cordoba itself—that the caption could just as easily read: "A militiaman slips and falls while training for action."[15] The value of the photograph lies in its claim to bear witness to the very instant of death—yet nothing in the image itself attests to that fact irrefutably.

Without a caption or title or some immediate context, a photograph has little power as a reproductive representation. With one, it retains the same capacity, the same power to mislead us as readily as any generative image: photographers working for William Randolph Hearst's *New York Journal* routinely altered file photographs and sent them to engravers, claiming these were legitimate shots of the unidentified victims of particularly grisly murders.[16] And Alexander Gardner's Civil War images that claimed to bear witness to the dead littering the battlefield at Gettysburg were actually rearranged by the photographer or his assistant: the body of "a dead Rebel sharpshooter" that appears in one image also turns up in a photograph of a "dead Union sharpshooter."[17] Presumably, Gardner believed that the dead, like Tolstoy's happy families, all seem alike in their own way.

The Long Dream of Representation

The problem is that, while reproductive and mimetic images can seem as if the world were simply captured on celluloid or canvas—in the way that Canaletto's meticulous renderings of Venice generated speculation that he actually used a camera obscura to trace their outlines—we cannot rely on them as reliable guides to the world around us. Try to assume a single stance to replicate the same perspective assumed by the artist, and you'll discover that Canaletto combined two or more viewpoints, changed the heights of the buildings and falsified the line of the horizon to create more harmonious compositions.[18] Which takes us back to Aristotle's old line about it being better to paint a hind without horns than to paint it "inartistically."[19] Sometimes we can convey the sense or feeling or conceptual core of an object or person infinitely better by *not* striving to represent it simply, completely, as it appears to us.

We can explain the other epistemological problem lying in wait by looking at two key words: *re-* productive, *re-* presentation. The subject is not being produced or presented to us firsthand; it is brought *back* before us for the second time, in a form that stands for the original. Some agent, some hand is acting the conduit for that *re-* production and therein lies the whole, nasty epistemological mess. The photo finish in a Triple Crown race, a recording of your step-grandmother's voice, a video of a car crash, all presume to bring us the things themselves, as they really are. Instead, they bring us the subject, arrested in time, framed within a static context, extracted from a stream of events, recorded at a specific angle, in a particular light, on a distinct type of stock, using a technology which may, like early experiments in Technicolor, turn vivid reds to muddy browns or normal skin to ash.[20]

The long dream of representation has been to somehow package reality tidily, in discrete portions small enough to be embraced or consumed. The long dream of representation has been to somehow parse existence and consciousness and knowledge and segment it into bite-sized chunks. The long dream of representation has been to deliver all things as they are, without bias or human intervention. The long dream of representation is, unfortunately, a chimera, a fiction, because representation is about filtering, selection, diminution: we use representations to filter and parse and reduce the world around us into bite-sized chunks because we cannot be omnipresent or omniscient and we rely on other minds and perspectives to filter for us what we cannot ourselves personally embrace.

Where, we might ask, is the point at which an image loses its representational purity, its reliability, its status as a testament to the pure existence of its subject? In the case of the infamous *National Geographic* cover, was it at the moment when, to use the words of one editor, someone "retroactively repositioned" the pyramids to make them into a more photogenic composition?[21] Or was it when the photographer waited for, or perhaps even hired, a camel rider to position man and animal at a particularly fortuitous spot in the frame, making for a picturesque scene? As Robert Brandt, managing editor of *Newsday* notes: "The potential for tomfoolery lurks no more dangerously at the computerized graphics and pagination device than in the low-tech equipment of the past."[22] Of course, the tomfoolery may be more difficult to spot. Although traditional approaches to photographic retouching and repositioning involved photographing the second-generation print to produce a seamless and real-seeming third-generation negative, some degradation in the sharpness of the image, some enlargement in its grain was often detectable—although these often worked to further smooth the image and conceal minor discontinuities even more (*RE*, 183). Even the existence of negatives—an "original" image where the generative object has none—doesn't provide us with any infallible link to the real thing, since retouched prints have usually been photographed a third time, producing a negative of a highly doctored image. The whole value of a photograph as a means of recording for posterity an instant arrested, fixed absolutely in time and space, has been the dangerously seductive secret to its value.

Having been weaned on the documentary value of reproductive representations, however, even skeptics like William J. Mitchell in his invaluable *The Reconfigured Eye: Visual Truth in the Post-Photographic Era* (1992) scrabble for the visual equivalent of a litmus test, some indicator to assess the congruence between the image and the world in which it is ostensibly set:

> The more information there is in an image, the harder it is to alter without introducing detectable inconsistencies . . . Furthermore, the difficulty of convincing alteration grows exponentially with the variety of types of visual evidence present . . . A photographic manipulator, like a dissembler who weaves a tangled web of lies and eventually trips himself up, is likely to be caught by some subtle, overlooked inconsistency. (*RE*, 37)

Certainly, the old bags of tricks may not have involved exactly seamless sleights of hand. Even William Roscoe Thayer, Walt Whitman's biographer, was highly skeptical of the veracity of the butterfly that just happened to flutter into the photographer's studio and perch photogenically on the poet's finger. He later noted, despite Whitman's protests that he had "always had the knack of attracting birds and butterflies and other wild creatures," that it was highly unlikely that the butterfly would even be alive on a day when Whitman was posing in a thick wool jacket in a closed studio (*RE*, 196). Photographers are fond of pointing out the same kinds of visual discontinuities Mitchell describes in shots like the Warren Commission photographs of Lee Harvey Oswald posing with his mail-order Mannlicher-Carcano: the shadow under Oswald's nose has been cast by sunlight from an angle different from the one casting shadows over the rifle he brandishes; Oswald himself stands at an angle that challenges every understanding of Newtonian physics.[23] But these assemblages that pass for original, singular photographs are no more implicitly faked than their nineteenth-century counterparts by photographers like Camille Silvy. Today, with the wide availability of photo libraries on CD-ROM offering millions of high-resolution images—not to mention the possibility of generating high-resolution scans from photographs and drawings for retouching, and the growing sophistication of retouching tools—puts an end to our ability to believe that reproductive objects are mere records of scenes, sounds, and events:

Does [digital imaging] mean we will no longer believe in the truth of the photographic images we see in our newspapers or on our desks? The problem with such a question is that traditional photographs—the ones our culture has always put so much trust in—have never been "true" in the first place. Photographers intervene in every photograph they make . . . by enhancing, suppressing, cropping the finished print in the darkroom; and finally, by adding captions and other contextual elements to the image to anchor some potential meanings and discourage others . . . the production of any and every photograph involves some or all of these practices of manipulation. In short, the absence of truth is an inescapable fact of photographic life.[24]

Beyond Mimesis, Back to Mimesis

> If [the poet] describes the impossible, he is guilty of an error; but the error may be justified, if the end of the art be thereby attained—if, that is, the effect of this or any other part of the poem is thus rendered more striking . . .
> . . . not to know that a hind has no horns is a less serious matter than to paint it inartistically.
>
> —Aristotle, *Poetics*

Once you let go of the truth value of reproductive technologies, the epistemological horrors subside, and the utilitarian value of generative technologies becomes clear. With tools that enable us to depict with photographic realism and digital precision what a completed building would look like—and VR technologies can enable us to even take a virtual walk down its corridors—we can make informed, fine-tuned decisions about the buildings we'll inhabit, the cars we'll drive, the look and feel of a kitchen, a corridor, a corporate boardroom. Generative technologies can give us what looks like a reproductive representation of a subject that does not yet exist. Or ways of seeing subjects impossible through our restricted, binocular vision, what Fred Ritchin calls "hyper-photography":

> Conventions such as one-point perspective may be seen as far too simplistic, and multiple points of view—physically and even psychologically—will be possible . . . One may also see in three or more dimensions, or scan a scene, asking for supplementary visuals or other kinds of information.[25]

Of course, what seems like generative sleights of hand are perfectly possible with older, mimetic forms of art. The tradition of simultaneously representing multiple perspectives of a single object or land-scape did not begin and end with cubism. All we need do is look back at the conflated perspectives of Canaletto's canvases or at David Hockney's recent paintings, "The Road Across the Wolds" or "The Road to York Through Sledmere," which portray the con-stantly shifting perspective one would enjoy from a car moving across the Yorkshire wolds, all fixed on a single canvas.[26] The differ-ence between mimetic and generative media is that generative media can do fluidly, seamlessly, and easily what mimetic tools ren-der in static, fixed, cumbersome ways.

What is new about generative media is not, as we have seen, the relationship between subject and object (remember Canaletto's Venice that never was), nor its monolithic nature (think of dyna-mism, cubism, or Hockney's most recent works), nor even its lack of linearity. What is revolutionary about generative technology is its capacity to represent perspectives on the things around us that can potentially morph into complementary or even mutually exclusive takes on successive encounters with it, without its beholders neces-sarily bringing anything new to the text.

I can plunge into Balzac's *Cousine Bette*, the book Zola champi-oned in *The Experimental Novel* as a behaviorist experiment and see the novel as a chronicle of the destruction wrought by Baron Hulot's seemingly bottomless lust. Rereading it years later, I can decide the crux of the narrative is about the power of pettiness, jealousy, and vengeance, particularly Cousine Bette's. But what is mutable is my interpretation, the bundle of biases, expectations, and experiences I bring to the novel. Yet when a reader encounters the mother in, for example, the hypertext short story *I have Said Nothing* as first a sympathetic, savior figure, and in subsequent readings as a margin-ally punitive one, the reader's shifts in perspective follow actual, physical changes in versions of the text she navigates through.

With such fluidity and multiplicity possible, generative media rep-resents something of an escape hatch for certain kinds of writing. Avant-garde fiction has been alternately playing with and chafing against the confines of the printed page since Sterne's day, but some genres of argumentative writing have been more recently physically challenged by print—hobbled, hampered, and turned into unwit-ting parodies of themselves. Philosophers, sociologists, anthropolo-gists, and proponents of critical studies have long ago abandoned the fixed, objectivist perspective on the world. Yet the conventions ruling print publication nearly inevitably corner them into behaving like covert objectivists. Social constructivists, for instance, begin writing about the interpretive flexibility of physical artifacts and the ways in which anyone can always dredge up fresh, equally compelling interpretations for them—and conclude with a singular observation, generally that their colleagues have latched onto too few, or even wrong, interpretations. Closure and conclusiveness, not surprisingly, are the most difficult aspects of print to duck, regardless of genre. Writing, as Plato implied in *Phaedrus*, is about controlling meaning beyond the reach of human voice and gesture.[27] Print, as Bruno Latour argues, is about the power to act at a distance, requiring powerful and conclusive explanations that feature

> a general feeling of strength, economy, and aesthetic satisfaction: the one element may "replace," "represent," "stand for" all the others, which are in effect made secondary, deducible, subservient, or negligible . . . An explanation becomes more powerful by relating more elements . . . to a single element.[28]

If we think of the socioeconomic forces that shaped print throughout its history, this outcome is hardly surprising. Writing and print have long represented ways of exerting control over a public mostly beyond the reach of the writer's voice, from Plato's *Phaedrus* on down. But as any seasoned researcher will tell you, the more texts and evidence you slog through and the closer you draw to the vortex of any project, the more hypotheses bloom and the less certain you tend to be about the total veracity or verifiability of any singular explanation. Relativism, our pluralistic society, the proliferation of knowledge and its greater accessibility, all nudge us closer to a world where digital technologies like hypertext might enable us to do more justice to our worldviews than anything possible within the confines of our current print conventions.

David Kolb's *Socrates in the Labyrinth*, a collection of hypertext essays, is currently the most comprehensive exploration of the potential interplay between generative technologies and philosophical argument. Among the more radical possibilities Kolb explores in hypertext philosophy are:

- recording dialogues and counterarguments
- making an argument discredit itself through proliferation
- dialogue without drama: regions in a landscape, not rival position papers

- landscapes with many rival surveys but no overview
- keeping accessible earlier options
- complexifying the connections between universal and the particular
- endless qualification or questioning or regresses of metalevels
- offering thematic unity or unities but without a line
- recontextualizing an argument or assertion, reusing pieces in new ways
- infiltrating, disrupting, or mixing the discourse(s) surrounding an argument[29]

Readers nosing their way through a philosophical hypertext might encounter paths representing mutually exclusive arguments, an expanding web of regions mounding questions upon questions, or thickets of commentary on commentary, following traditions familiar to us from Talmudic scholarship.[30]

Writers struggling to crystallize a perspective through written argument often fret that their readers may be baffled by the premises, disagree with the conclusions, and quibble over the validity of every assertion. Writers using Kolb's argumentative structures—if they can even still be called "argumentative"—won't know whether their readers will happen upon all the branching possibilities, which argumentative lines strike them as more compelling than others, or even if their audiences decide to oscillate between two or twelve possible fixes on the subject. Argument need not be a matter of swaying opinions from outside our shouting range or beyond the grave; it can simply act as a venue for spreading out a full range of interpretive possibilities, allowing the reader to either reach, reject, or suspend conclusions.

Of course, hypertext arguments can still be palpably authored, perhaps by a *deus absconditis* rather than a *deus ex machina,* a god who throws down all the characters and essential dilemmas and then buggers off without bothering to hang around for the results. An argumentative text that unspools into ten or fifteen mutually exclusive hypotheses with different assertions based on the same pool of evidence is not evidence for the death or necessarily even the diminution of the author, any more than it represents total freedom for the reader, all of which have been claimed about hyptertext. Like the pronouncements about the daguerreotype and the motion picture, these claims are also simply part of the same pre-Lapsarian longing for representations that deliver the world to us in manageable chunks without detectable limits or the usual greasy fingerprints.

Hypertext, with its relative lack of conventions, has a fluidity with which writers can order and reorder information without the static physical limits and closure of the printed word, enabling us to more easily convey, for example, the interpretive flexibility of socially constructed artifacts, like Langdon Winner's famous bridges in "Do Artifacts Have Politics?" After making a muscular and now famous argument for implicit social and political agendas in an artifact as mundane as a parkway bridge, Winner fixes onto a single interpretation—Robert Moses, builder of the New York parkway network, engineered his own racist and elitist agendas into the height of its bridges saving his parks, presumably, from stampeding hordes of unwashed inner-city immigrants and minorities.[31] Conclusion reached. Yet if technological artifacts are texts that can be read and interpreted like *David Copperfield* or *Waiting for Godot,* as Winner suggests, then surely accepting as your ultimate and only conclusion a reading based on a remark made by a disgruntled former Moses lieutenant may not be the most reliable interpretation.[32] Why not mention that Moses designed his rustic stone bridges and forest-hemmed parkways to capitalize on the pleasures of country drives, which led him to prohibit commercial traffic on the entire parkway system?[33] Or note that the height of Long Island's stone parkway bridges merely conforms to the same standards as the Lincoln, Holland, and Brooklyn-Battery tunnels, which collectively accommodate hundreds of buses every day? Any reader remotely au fait with the history of Greater New York could point out that the public buses Moses would have known when he was designing his parkway system were far from the lofty, wide versions we ride in today. And New Yorkers know that, if Moses' parkway bridges restricted the residents of Brooklyn, the Bronx, and Queens' poorer pockets from riding buses to the system of beaches and parks he created on Long Island, they also prevented them from visiting the none-too-swish areas connected by the Cross Island Parkway.[34] As the social constructivists like to say: it could always be otherwise.

"Providing an explanation . . . is inherently good; thus accusing someone of providing no explanation puts an end to the dispute," writes Latour in "The Politics of Explanation":

> the opponent is just story-telling and may be stopped by a simple question like "so what?"; to answer the "so what?" question entails proving that he or she is doing *more* than just telling stories, that he or she is really offering some explanation.[35]

Perhaps this is a modest claim, that generative media have the capacity to provide a range of perspectives we're invited to occupy, where the representative and mimetic conventionally tend to funnel our perceptions into singular views. The current possibilities offered by generative media may well be more modest than this, particularly since few writers necessarily perceive the world "multiply," as Jay Bolter suggests of the hypertext aesthetic.[36] Possibly still fewer, even if they knew how to represent things multiply, would surrender their fleeting opportunity to embed their own convictions in our thoughts in favor of making, say, a half-dozen positions equally tenable, including some they themselves may find personally unpalatable. And it also takes time—decades, centuries—for us to be molded by the tools we ourselves fashion. Yet there are already little pockets of writers out there even now, already impatient with the limitations of our mimetic and reproductive tools, some before they've so much as glimpsed their first hyptertext.[37]

What can we expect? Probably representations that blur the boundaries between narrative and argument, writings that are not intended as tools to act at a distance but as maps providing alternative views of the same terrain, an aesthetic that may well value comprehensiveness over conclusiveness, rich complexity and even cognitive overload over singularity and elegant concision. Mulling over the state of writing and representation in the social sciences over a decade ago, Latour arrived at a similar conclusion:

> The stylistic conclusion is that we have to write stories that do not start with a framework but end up with local and provisional variations of scale. The achievement of such stories is a new relationship between historical detail and the grand picture. Since the latter is produced by the former, the reader will always want *more details*, not less, and will never wish to leave details in favor of getting at the general trend. This also means that stories which ignore cause and effect, responsibilities, and accusations, will be unfit for the normal mode of denunciation, exposition, and unveiling.[38]

Perhaps this is the "new realism" that Michael Joyce's student, Josh Lechner, felt hypertext might represent, more of an approximation of William James's "buzzin', bloomin' confusion" of the world as it swarms our senses and thoughts than the singularity and purposiveness of the world organized by print.[39] Perhaps this new mode of nonargumentative argumentative electronically mediated writing merely reflects a world in which we feel we've already seen, heard, read, and experienced it all, virtually. Perhaps we now feel conclusiveness and singularity represent limitations in an era where change is rapid and almost immediate obsolescence inevitable. This much is certain: it's even becoming easier to avoid reaching singular conclusions in our old, reliable mimetic media—like print, like now.

Notes

The first two epigraphs are from Fred Ritchin, *In Our Own Image: The Coming Revolution in Photography* (New York: Aperture, 1990), 2, 15; the third epigraph is quoted in Geoffrey Batchen, "Phantasm, Digital Imaging and the Death of Photography," in *Metamorphoses: Photography in the Electronic Age* (New York: Aperture, 1994), 47.

1. Sir Joshua Reynolds, "Discourse XIV," *Discourses* (New York, 1928), 240–41.

2. Edgar Allan Poe, "The Daguerreotype," in Alan Trachtenberg, ed., *Classical Essays on Photography* (New Haven, Conn.: Leetes Island Books, 1980), 37–38.

3. Pliny, *Natural History* XXXV, quoted in E. H. Gombrich, *Art and Illusion: A Study in the Psychology of Pictorial Representation*, 2d ed. (Princeton, N.J.: Princeton University Press, 1962), 206.

4. Gombrich, *Art and Illusion*, 11.

5. Benoit Mandelbrot, *The Fractal Geometry of Nature* (New York: Freeman, 1983).

6. Ritchin, *Our Own Image*, 17.

7. Ibid., 9–13.

8. Mark Haworth-Booth, introduction to *Metamorphoses*, 2.

9. William J. Mitchell, *The Reconfigured Eye: Visual Truth in the Post-Photographic Era* (Cambridge: MIT Press, 1992), 183. Hereafter *RE*, cited in the text.

10. Louis D. Giannetti, *Understanding Movies*, 2d ed. (Englewood Cliffs, N.J.: Prentice-Hall, 1976), 384.

11. Karal Ann Marling and John Wetenhall, *Iwo Jima: Memories and the American Hero* (Cambridge: Harvard University Press, 1991), quoted in Mitchell, *Reconfigured Eye*, 43.

12. Penelope Niven, *Steichen: A Biography* (New York: Potter, 1997), 48.

13. Walter Benjamin, "A Short History of Photography," in *Classical Essays*, ed. Trachtenberg, 215.

14. Roland Barthes, "The Photographic Message," in *Image-Music-Text*, trans. Stephen Heath (New York: Hill & Wang, 1977), 26.

15. Phillip Knightley, *The First Casualty* (New York: Harcourt Brace Jovanovich, 1975), 209–12.

16. Harry J. Coleman, *Give Us a Little Smile, Baby* (New York: Dutton, 1943), quoted in Ritchin, *Our Own Image*, 210.

17. Frederick Ray, "The Case of the Rearranged Corpse," *Civil War Times*, October 1961; quoted in Mitchell, *Reconfigured Eye*, 43.

18. Violet Pemberton-Pigott, "The Development of Canaletto's Painting Technique," in *Canaletto*, ed. Katherine Baetjer and J. G. Links (New York: Metropolitan Museum of Art, 1989), quoted in Mitchell, *Reconfigured Eye*, 188.

74   ADRIFT IN THE TECHNOLOGICAL MATRIX

19. Aristotle, *Poetics*, trans. S. H. Butcher in *Critical Theory Since Plato*, ed. Hazard Adams (New York: Harcourt Brace Jovanovich, 1971), 64.

20. As Mitchell has noted in *Reconfigured Eye* (188ff.), photo finishes in thoroughbred racing are actually wide-angle images on film pulled past a vertical line that stands in for the finish line, representing "the changing state of a narrow field of view over the time that it takes for several horses to pass through it" (251), itself not the fail-safe procedure it originally seems. More striking still is the significant disparity between photographs and videos of races and the "reality" represented by the official track clock. Take, for example, the time recorded for Secretariat's 1973 Preakness win by Pimlico track officials and that established through frame-by-frame analysis of the official video of the race. A full :001 difference exists between the recorded clock time of 1:54:2 and the 1:53:2 recorded on the video, verified by frame-by-frame analysis with the clock times recorded for the race's other entrants. The first is a fast time; the second would have toppled the existing Preakness record. Maryland thoroughbred racing officials stuck by the time recorded on the clock; *The Daily Racing Form*, the trackside bible, attributes the record to Secretariat (later tied by Tank's Prospect in 1985 and Louis Quatorze in 1996).

21. Quoted in Ritchin, *Our Own Image*, 14–15.

22. Robert F. Brandt, "Technology Changes, Ethics Don't," *Presstime* (December 1987), 32.

23. See Mark Lane, *Rush to Judgment* (New York: Holt, Rinehart & Winston, 1966), 356–62, where Lane also documents retouching and cropping that significantly altered the representations of the photographs that appeared in *Life*, *The New York Times*, *Newsweek*, and *The Detroit Free Press*.

24. Batchen, "Phantasm," 48.

25. Ritchin, *Our Own Image*, 124.

26. Hockney's paintings were on display at the Museum of Fine Arts, Boston, April–July 1998.

27. Plato, *Phaedrus*, trans. Walter Hamilton (London: Penguin Books, 1973), 96–99.

28. Bruno Latour, "The Politics of Explanation: An Alternative," in *Knowledge and Reflexivity: New Frontiers in the Sociology of Knowledge*, ed. Steve Woolgar (London: Sage, 1988), 158.

29. David Kolb, "Socrates in the Labyrinth," in *Socrates in the Labyrinth: Hypertext, Argument, Philosophy* (Cambridge, Mass.: Eastgate Systems, 1994), "some phil. HT actions."

30. Ibid.

31. Langdon Winner, "Do Artifacts Have Politics?" in *The Whale and the Reactor: A Search for Limits in an Age of High Technology* (Chicago: University of Chicago Press, 1986), 23.

32. Robert Caro, *The Power Broker: Robert Moses and the Fall of New York* (New York: Vintage Books, 1974), 318–19.

33. Ibid., 162.

34. J. Yellowlees Douglas, "Will the Most Reflexive Relativist Please Stand Up: Hypertext, Argument, and Relativism," in *Page to Screen: Taking Literacy into the Electronic Era*, ed. Ilana Snyder (New York: Routledge, 1997), 150–51.

35. Latour, "Politics of Explanation," 156.

36. J. David Bolter, *Writing Space: The Computer, Hypertext, and the History of Writing* (Hillsdale, N.J.: Erlbaum, 1991), 142–46.

37. See, for example, Kolb's account of his work prior to encountering hypertext in "Socrates Apology," *Seulemonde* (University of South Florida, Spring 1995), "Eugene, OR."

38. Latour, "Politics of Explanation," 174.

39. Michael Joyce, "New Stories for New Readers: Contour, Coherence, and Constructive Hypertext," in *Page to Screen*, ed. Snyder, 174.

"New World" Citizenship in the Cyberspatial Frontier

Catherine Gouge
West Virginia University

> To the frontier the American intellect owes striking characteristics. That coarseness of strength combined with acuteness and inquisitiveness; that practical, inventive turn of mind, quick to find expedients; that masterful grasp of material things, lacking in the artistic but powerful to effect great ends; that restless, nervous energy; that dominant individualism, working for good and evil, and withal that buoyancy and exuberance that comes from freedom—these are the traits of the frontier, or traits called out elsewhere because of the existence of the frontier.
> —Frederick Jackson Turner, *The Frontier in American History*

AS Frederick Jackson Turner argued in 1893, the concept of the frontier has been integral to the formation of an American national identity, a national identity that is predicated on intellectual as well as material control over the natural world. Turner's rhetoric, in fact, emphasizes a virtual frontier "of mind," a conceptual frontier that is dependent upon its ability to call forth an energy which enables a "masterful grasp of material things." The idealized "masterful grasp" of which Turner writes—implicitly aligned with an objective, "scientific" intellect, "lacking in the artistic but powerful to effect great ends"—veils the colonial desire inherent in frontier exploration. Such a representation of the promise of the frontier serves, furthermore, to mystify the seamless imbrication of colonial desire and frontier "freedom" in the production of the "practical, inventive . . . dominant individualism" which Turner claims is "called out" by the American frontier.

Throughout Turner's discussion of the agricultural promise of the early American West, he implies that the concept of the frontier—as an open, "free," and formless space—defines Americans by calling forth specific "intellectual" characteristics which are unique to our attempts to explore and develop frontier space. In so doing, Turner suggests that the idea of the frontier calls out symbolic, virtual traits which are not merely "traits of the frontier" in the early American West; they are, more generally, as he also writes, "traits called out elsewhere because of the existence of the frontier"[1] as an imaginary—and, thus, immaterial—space. If in its constitutive potential the concept of the frontier begs our participation and, thus, functions to define us through our desire for and attempts to secure the "masterful grasp" of which Turner writes, we ought to consider the ways in which our subjectivities are mediated by our participation in frontier ventures. To this end, we might question the ways in which our subjectivities are "remediated"[2] by our activities in the cyberspatial frontier and the ways in which agency in representations of cyberspace is contested and completely reinscribed to fit a paradigmatic—imaginary—notion of self as corporate manager.

Invoking Turner's rhetoric to argue for the budding sovereignty of what he calls the Digital Nation, Jon Katz wrote in 1996 that "the Internet is still a wild frontier" and compared "digital" users to those who colonized North America and the Communications Decency Act of 1996 to the Stamp Act of 1765:

> Like the colonists, the online community saw the law as an arrogant act by an alien entity seeking to force its will on a new world that it had lost any moral right to control . . . If the Stamp Act marked a turning point in the colonies' relationship with England, the CDA did the same for the digital world, giving credence to the notion of the birth of a Digital Nation. The CDA's passage and the Digital Nation's reaction to it showed that the digital world was creating not a radically new value system, but that it was now the champion for a venerable old one: the notion of individual liberty.[3]

Katz's appropriation of Turner's rhetoric underscores the current trend in contemporary popular culture to invest cyberspace with the rhetoric and putative value of the frontier, while repressing the historical experience of the frontier that propels such rhetoric. By aligning cyberspace and the promise of this "new" frontier with traditional values of "individual liberty," Katz privileges a technology-is-destiny myth which masks the complex antagonisms inherent in any frontier venture. Building on the discursive formulations circulated during the "space race," such rhetoric reinscribes the frontier ideology that makes technological sophistication one measure of empowered American citizenship. Accordingly, such representations of the cyber frontier function to constitute us as functions of capital.

The frontier metaphor was crucial to the formation of particular frontierist attitudes toward cyberspace. Advertisements for computer technologies of the 1990s invited our participation in the cyberspatial frontier, encouraging us, thus, to reimagine our participation in figurative frontier spaces as a way of reinforcing our power as "productive" citizens of a nation-state premised on a frontierist ideology. Such narratives help to empty the subject of meaning and treat the subject as a placeholder for a fiction of power. This fiction erroneously imagines that the powerful citizen-consumer is "complete" so that it can insist that we cannot be complete or powerful without economic agency, and we cannot have economic agency without the newest and "best" technology. If our subjectivities in the cyberspatial frontier are effects of technology and this technologically "empowered" subjectivity is dependent upon and necessary for economic prosperity, then the powerful and productive citizen-consumer—or so the logic goes—can be understood most accurately as the economically powerful subject with the latest, and thus most "powerful," technology. This is reflected in many advertisements for computer-oriented materials in which the emptying out of interiority is appropriated by corporate types. According to these ads, our economic agency must be always beyond our grasp because it is predicated on the same notions of planned obsolescence that allow companies to sell computers, among many other products.

Therefore, while Katz and others would like to suggest that cyberspace is a democratic frontier space characterized by Enlightenment ideals in which everyone is equal, I would like to challenge the ideology that propels such a recycling of frontier metaphors. The cyberspatial frontier, like other postoriginary[4] frontier spaces, will not, in fact, fulfill our collective fantasies of wholeness, competence, and power. It will not, furthermore, allow us to transcend our fragmented, insecure postmodern bodies. Rather, with our participation, it will continue to reproduce us as effects of a symbolic network defined over and against the "inadequate" real, material world, coded as static and immobile. Precisely as it materializes as a frontier—as we constitute it and represent it in terms of its opposition—cyberspace further constitutes our subjectivities in its own frontierist image. We must, therefore, make ourselves aware of the ways in which others—primarily those marketing various new technologies—are reinscribing notions of American citizenship by defining for us the "consensual hallucination"[5] that is cyberspace in frontierist terms.

In their attempts to invest in the supposedly invaluable promise of the cyberspatial frontier, many advertisements for computer-oriented materials propose cyberspace as both a prosthesis that "adds to" or embellishes our real life experiences and a virtual, complete realm—a simulation in which we can "transcend" the physical world.[6] Part of what such ads are selling us, then, is a unified, enhanced, virtually embodied version of our own subjectivity and the notion that we can occupy or "control" any subject position and body we desire in cyberspace. In short, we are being sold a notion of unified subjectivity and agency that has never existed and can never exist for any of us in the real world. These ads, then, illustrate the ways in which those marketing such products are themselves implicated in a colonizing ideology which dictates our desire as a "masterful grasp of material things" and which defines the key to "whole," productive citizenship as being the most powerful consumer. And if we fail to recognize the ways in which our desire is being overdetermined by such a frontierist ideology, backed by a whole tradition of metaphysical philosophy, we overlook a significant paradox: technology cannot be both a prosthesis and an essential means to recovering an "original identity."[7]

We are all marked by the anxiety of not being technologically sophisticated "enough," a concept which reinscribes a cutthroat, Hobbesian view of the economic world. This anxiety is, furthermore, reinforced by many advertisements for various computer-oriented materials that subscribe to the notion that computer technology can be used to enhance our sense of power over our lives. In the service of such a logic, we are made to believe that if we use this cyber technology effectively we can have a sense of power through control over our virtual and real socioeconomic resources. A 1997 Creative ad thus asks, above an image of a suited white man trapped inside his computer monitor: "Feeling boxed in by your PC's multimedia capabilities?" Assuming an affirmative reply the ad responds: "It's time to upgrade your PC with Creative!" and offers an image of the same man, now depicted as clownishly enthusiastic, emerging from a colorful, "wild" cyberspatial psychedelic background window. A series of AST ads similarly imply that our desired economic agency is predicated on technological adeptness. They announce their

motherboard's FlexChassis design which "allows easy access to the
system's major components in seconds." "Since even the best technology will one day need upgrading," the ad tells us, "AST is here
to help you do it right."[8] These ads, then, reproduce a larger logic
of technological angst that depends on constantly reinforcing—and,
the other half of the dialectic, undermining—our illusions of control over the very technologies that intervene in and reinscribe our
identities. In so doing, the ads redefine a problematic, neurotic,
"real" identity as economic agency—always imperfectly, of course.
The irony, then, is that even as virtual technologies "demonstrate"
that they are essential to complete an inadequate identity and sense
of self, they promote the fiction of an integrated, coherent, and
powerful (or all-powerful) self—an always symbolic fiction that re-
cedes whenever we try to articulate it.

The effects of our interaction with such postoriginary frontiers
can present themselves in many different ways. Robert Markley
writes that "the experience of virtual reality is intended to send us
back to real reality with a heightened appreciation of the exploitabil-
ity of our environment."[9] This assertion points to at least a few different potential effects of cyber experiences on the user. That is, the
cyber participant might exit virtual realities with the emotional and
intellectual response for which Katherine Hayles calls, namely, "remember[ing] the fragility of a material world that cannot be replaced."[10] Or, the user might exit with a "heightened" desire to capitalize on our potential to control "reality, nature, and especially . . .
the unruly, gender- and race-marked, essentially mortal body."[11] In
an effort to reinvest in fictions of technologically mediated agency
and power, these ads generally appeal to four frontierist schemes:
that cyberspace is, following in the wake of the exploration of "outer
space," the "new" frontier; that we must view our efforts to tame
this frontier as a war in which the computer is our most powerful
weapon; that once we have tamed it, we can sit back and exploit the
resources of this new civil, democratic territory; and that in this
"equal" space, we can, ironically, luxuriate in our newfound economic power. Indeed, in order to feed anxieties that we are not technological enough, the images of the computer in advertisements
draw on that aspect of our cultural identity that distrusts and/or
fears technology. So, while the rhetoric of Katz and others is
couched in the language of American individualism, the practice of
cyberspace is, at once, militaristic (command, control, communication) and consumerist. The ads are trying to sell us an ideology of

consumerism as well as hardware and software. Therefore, our supposed "agency" in these ads is defined by the "stuff" we can purchase. In short, such ads reinscribe our power as citizens as an effect of our existence as consumers.

For many proponents, the key to masking such attempts to reinscribe our citizenship lies in their representations of our journey into cyberspace as one constituted by American frontierist metaphors of escape as well as colonization. These metaphors of escape and colonization create different conceptual views of cyberspace; however, because we are trained to see language as ornamental rather than as constitutive, we end up with confused, contradictory images of cyberspace because we do not recognize that we have constructed contested and possibly incompatible images of virtuality and identity.[12] Thus, while this "modern" New World or Digital Nation is represented by Katz and others as driven and peopled by those who are defined by their quintessentially American search for freedom from the "constraints on individual expression" placed on Digital Nationals by the nondigital real world, this "real world" is constructed by Katz and others as a back formation. The real world is defined as the antithesis of the Digital Nation that has no "real" existence prior to its fantastic construction. We must recognize, therefore, the ways in which the cyber frontier is driven and peopled by those who are motivated to exploit this "new" and infinite space's money-making potential.[13]The danger, then, is at least twofold: narratives of cyber technologies foster a retreat into a frontierist hyperindividualism that Katz invokes (in a much less critical way than Turner) and serve to provoke a self-propelling desire for an unmediated experience of the real that can be satisfied only by increasingly sophisticated technologies that get us "back" to the real, a nature that is always and already being reproduced by technology. The very logic of the cyberspatial frontier narrative, then, betrays its supposedly progressive agenda. It is not, in fact, a narrative that will take us *into* the future, a temporal beyond; rather, it is one that is designed to make us desire a mediated *return to* the real, material world. By creating such a desire to return to the real through cyberspace, such narratives define our desire as desire for the real world mediated by the simulation. Such narratives attempt to conflate the simulation with the real and thus suggest that our best chance at economic agency is through a masterful grasp of the simulated real, the cyberspatial real. However, because we can never be technologically sophisticated enough, this is a grasp, of course, that is—like the frontier—always and already beyond our reach.

Most colonial projects begin with the desire for economic control and superiority and are predicated on the assumption that the colonized are inferior because they lack the technological sophistication that the colonizers see as integral to their own cultural identity.[14] Michael Adas notes that historically "scientific and technological measures of human worth and potential dominated European thinking . . . [and] provided key components of the civilizing-mission ideology that both justified Europe's global hegemony and vitally influenced the ways in which European power was exercised."[15] This view of "natives" as technologically inferior is, in part, an externalization of "our" worries and anxieties that "we" are not technologically sophisticated enough. In response to such imagined anxiety, Pioneer—whose name evokes the frontier metaphor—directs readers of *Wired* magazine to "link into the brave new digital world." The accompanying graphic depicts, as small print at the bottom of the page explains, "one possible scene in the future as imagined by Pioneer."[16] In this depiction appear two models of the globe—one is a picture of earth from outer space, superimposed with inch-wide squares illustrating the various images of corporeal activity (kayaking, playing baseball, working on the computer) to which one might link through Pioneer technology; the other is filled with a "futuristic," ecstatically happy, white family (replete with father, mother, and two children) in their living room, thrown into relief by a TV/computer screen. The two globes are visually connected by the various cyberspatial vehicles Pioneer suggests will allow future happy, white families to conquer and civilize this brave new cyberspatial world: a satellite, broadcast cables, a digital cable, DVDs, and a computer network.

This "new" frontier, then, incorporates and remediates the technologies of the previous "final" frontier to suggest that computer technology, as cyberspatial vehicles, is what is needed to penetrate both the future and cyberspace.[17] This logic of remediation is further perpetuated by an ad for a Compaq laptop which advises, "Your PC is mission control . . . with seamless synchronization you've got everything you need until reentry." This ad constructs an idealized reader as one who privileges the masculinized power of control through penetration of the cyber frontier. On the opposing page, encircled by what appears to be a ground-up shot from the courtyard of the Watergate hotel, Compaq explains: "It Retrieves E-mail,

Accesses The Internet, And Connects You With Your PC. Because Sometimes You've Just Got To Contact The Mothership."[18] The function of conflating the spatial, maternal, and science-fiction metaphors in the rhetoric of the "Mothership" is to make outer space the fantasy which becomes actualized in cyberspace. "Contacting the Mothership" in cyberspace thus becomes the realization of an imaginary outer-spatial, frontierist conquest as well as a return to the womb. It becomes, in other words, both a manifestation of our technological accomplishment as well as a return to an imagined original innocence and wholeness.

A Toshiba ad, which frames its new laptop and optional Desk Station V Plus with the words "INTRODUCING A FLIGHT OF THE IMAGINATION. COMPLETE WITH LANDING GEAR,"[19] appropriates the traditional views of what the future in outer space will look like and exposes the cyber-frontierist ideology by suggesting that the creation of a unified, masculinized cyberspatial identity can take place only in our imaginations and cyberspace. The ad implies that the success of the "flight" made available to us through the technology requires our imagination and, more importantly, is defined by our imaginary capacities. The proverbial ship and landing gear, the ad suggests, are real; the trip is not. Furthermore, by defining cyber exploration in terms of the exploration of outer space, this Toshiba ad and others use one form of frontierist science-fiction imagery and rhetoric to overdetermine another. Such ads define one imaginary realm—cyberspace—with romanticized views of another—outer space. This ideological and virtual overdetermination pulls the apparently immaterial spatial quality of cyberspace "back" to the real and thus constructs and defines its materiality as masculine and, therefore, powerful. This recalls Theodore Roosevelt's promise that the frontier "will turn you from a weakling into a man."

Some ads use more obviously violent imagery to invoke notions of colonial angst in our attempts to seize the economic agency we supposedly deserve. Below the image of a laptop computer, for example, a Micron Electronics ad states: "Business is war. Bring a Howitzer."[20] Images of business as war foreground somewhat jingoistic expressions of "agency." In such representations, the goal becomes to have the strongest, most powerful computer-oriented weapons so that we can fight for and ideally own the mobility and completeness that comes with economic power. In reducing technology to weapons to suggest that subjectivity is an effect of capital, such ads draw our attention to the violence that is encoded in and often masked by frontier rhetoric. Other ads build on the concept of a society at war in cyberspace and propagate a contemporary version of the Manifest Destiny myth which insists that we must conquer the cyber frontier and exploit its infinite resources. This individualism is portrayed as equality of opportunity—a false equality in which the computer literate have a distinct advantage. The logic of such narratives thus depends on the fiction that by exercising our "individualism" we can create a democratic space in which everyone is equal—or, more insidiously, a place where opportunities are equal. Like America's original ideal of the melting pot whose goal was to forge one American national identity, those marketing cyber technologies would have us believe that our best shot at becoming members of this community is by assuming a unified, albeit virtual, cyber identity by surrendering to a logic of constant upgrades and, thus, achieving virtually something we cannot achieve in reality.[21] As a 1996 MCI ad directs: "Imagine a world in which everyone is equal, a world in which there is no gender, no race, no infirmity."[22] Aligned with infirmity, the markers of gender and race are here implicitly identified as bizarre emblems or manacles, chaining us to our racially embodied and gendered fates and histories and keeping us from the promise of the honorary white male "super-citizenship" which cyberspace offers us. This ideology of "equality" is, furthermore, ironically reminiscent of American frontier rhetoric in that it privileges downplaying and even erasing corporeal markers. Anyone can make it on the frontier, the ad suggests, even those "Others" who do not quite fit in the nonfrontier.

Katz and many of the ads, then, assume a neurotic, Freudian subject who desires autonomy and "completeness" and will spend "his" entire life working for them as long as he is provided with the illusion that he can actually have what he desires: coherence and power as a citizen through technological and economic superiority. In response to this, ads reinvest in the promise of economic agency for those who, by most standards, already have economic agency. In arguing for "a new kind of [masculinized, 'rational'] nation," which, although admittedly populated predominantly by men who are "richer, better educated, and disproportionately white" and "not representative of the population as a whole," Katz suggests that the cyberspatial frontier "celebrates the right of the individual to speak and be heard—one of the cornerstone ideas behind American media and democracy." "Like it or not," Katz writes, "this Digital Nation possesses all of the traits of groups that, throughout history, have eventually taken power. It has the education, the affluence, and the privilege that will create a political force that ultimately must be reckoned with."[23] With his assertion that the generally affluent and technologically advanced members of the Digital Nation are the superior "force to be reckoned with," Katz imposes a traditional reading of the American frontierist dynamic onto cyberspatial ventures and positions Digital Nationites as the modern colonizing force. The notion of the Digital Nation as refighting the Revolutionary War—a notion to which the ads documented earlier also subscribe—mystifies what is at stake in the digital revolution: individual citizenship in a real world American nation-state. However, by encouraging the public to conflate coherent citizenship, economic power, and technological adeptness, such narratives encourage the public to refigure their role in society as, ironically, antidemocratic. In so doing, such narratives encourage us to refigure our subjectivity as a function of capital and to disjoin power from meaningful labor, emphasizing the ways in which we are functions of capital and the ways in which such emptying out of interiority has been appropriated by corporate managers.

Consequently, the notion of technology's ability to function as an empowering civic prosthesis propagated by so many advertisements and proponents of computer-oriented technologies masks the economic conditions and motivations that characterize the exploration of this new frontier. It masks, in short, the proposition that those who do have access to computer technology should endeavor to exploit those without "full" access to the digital frontier ventures. Left unexamined, such an imperial ideological agenda is capable of recreating what it imagines. It not only is capable of reinvesting in a frontierist hierarchy of values which privileges economic and technological superiority as a measure of individual worth, but it is capable of suggesting that such privileges necessary for cohering identity are really choices we make as citizens of a "free" Digital Nation.[24]

However, as I suggest, the ads are effective only to the extent that they cannot completely deliver the technological and economic power they imply that we need. The "completeness" supposedly afforded by the technologies the ads support is always already marked by notions of planned obsolescence. The ads, thus, reproduce an always self-perpetuating desire for completeness by investing in our incompleteness. If we buy into the fictions these ads promote, we commit to buying our new and improved/empowered subjectivities on an installment plan—we can never be "complete" because, as

Žižek points out, wholeness is a projection of our symptoms. Furthermore, such completeness, as ads suggest in a typically frontierist gesture, must remain always beyond our reach so that computer companies can continue to sell computers.

In 1991 David Tomas wrote that new virtual reality technologies will ultimately enable users to "overthrow the sensorial and organic architecture of the human body, this by disembodying and reformatting its sensorium in powerful, computer-generated, digitized spaces."[25] The implication here is, of course, that the body is (as Edward Said argues about the colonized subject, somehow "backward") or unable to fit in[26] and that new computer technology will both provide us with an escape from the dead weight of our corporeal bodies and offer us a utopian, democratic, "free" frontier space in which our virtual, "whole" selves can be equal members of a cyber community. Furthermore, such narratives suggest that computer technology offers us a utopian space in which we can act out all sorts of desires—many of which would hardly qualify as "democratic." Cyberspace is, according to the logic of the narratives, merely a frontier settlement directed by the ideological pandects of real world hegemonies. Indeed, while attempting to suggest that cyber communities are places for individuals fleeing the constraints of the real world, these advertisements concurrently subordinate individual civic identities to a default citizenship: an implicitly masculinized, white, "whole" digital identity. An ad for GeoCities further reinforces the presumed corporeal insufficiency of those who do not fit into the white, male "super-citizenship" and appeals to those members of society who, it implies, are hopeless of having an equal voice outside cyberspatial parameters. In fact, the haphazard inclusivity of the list in the ad seems to appeal to our anxiety that we are lacking, hopeless eccentrics—that is, "genuine" "individuals" in a consumerist, mundane world. The ad calls out to

> Trekkies, Bookworms, Honky Tonkers, First Nighters, Einsteins, Bureaucrats, Undergrads, Jocks, Twerps, Supermoms, Groupies, Gurus, Motorheads, Foodies, Eurotrash, Tree Huggers, Big Spenders, Phreaks, Bohemians, Shiny Happy People, Gays, Head Bangers, Sun Worshipers, Couch Potatoes, Gamers, Taoists, Culture Vultures, Money Baggers, Ski Bums . . . **people like us.**[27]

The ad consoles these "hopeless eccentrics" by reassuring them that they can "fit in" to a community of others like themselves that will presumably fill their lack and make them whole: "Welcome to the ninth-most-visited site on the Web, GeoCities. 30+ neighborhoods of people like us who have 'homesteaded' Home Pages in the communities of their choice—Personal Home Pages—rich with content that allows us to find, and be found by, people who share our ideals, interests and passions." This ad, fairly quickly becoming an historical artifact, explicitly recalls the discourse of "homesteading" on the originary American frontier; furthermore, the notion of "homesteading Home Pages" speaks to an assumed need to overcome our respective material—physical, geographically limited—locations and suggests that "ideals," "interests," and "passions" require mediation by and through computer-oriented technologies to qualify as characteristics of individuality.

What the ads do not spell out, of course, is that our imaginary participation in the production of such technological and economic agency is indispensable and as soon as we begin producing we cease to have the potential for achieving the "masterful grasp" the frontier purportedly offers us. To participate, in other words, we have to imagine our corporeal lack and desire that it be repaired. We are constructed, therefore, as imaginary laborers who produce, ideologically, the very commodity—frontierist American citizenship—such ads purport to offer us as consumers, otherwise rendered immobile by our corporeal existence. An ad for Webtv in *Self* magazine deliberately reinforces our obsessions with our otherwise inert "lacking" bodies so that it may suggest that its Webtv technology is the answer to our corporeal laziness. It reads:

> Your butt.
> It's practically a fixation.
> You look at it in the mirror.
> You exercise it.
> You dress it, so it looks good.
> You pinch it, for firmness,
> like a melon.
> How can we help improve it?
> Sit on it, and find out.[28]

By suggesting that our corporealities require mediation through computer technology, this ad purports to give women a choice: use the technology to exploit the cyberspatial frontier or become one of the technologically and corporeally insufficient many who will be left behind by computer technology. The illusion of such a choice, among other things, suggests that in this most recent frontierist dynamic, we can choose to be a part of the technologically sophisticated, virtual, mobile frontier explorers or choose to be a part of the inactive, "unnaturally" lazy, implicitly colonized, masses. In this structure, stasis and corporeality are not only inhibiting, they are the wrong "choices," as undesirable as any historically subordinate class, race, gender, sexuality, or religion.

Some ads, however, do not map as neatly onto the power dynamic of traditional colonial narratives. An ad for Kingston technology, for example, positions readers as colonizers and technical supporters as native servants. With its suggestion that "you" had better buy their new, "superior" memory for your laptop, "unless you want to bring those tech support guys with you,"[29] Kingston positions the "tech support guys" as those who are routinely expected by those in power to serve their needs, work Gayatri Spivak terms "zero-work."[30] In this Kingston ad, the idealized cyber frontier dynamic is revealed to be that in which the "average" person (read: the aggressive corporate manager) is the powerful agent while those with sophisticated technological knowledge—knowledge of the "new" frontier—are the appropriated labor (like the Chinese in the building of the transcontinental railroad). This is a critical reversal of the usual colonial power dynamic whereby, as Adas points out, native inferiority is determined and defined by a person's lack of sophistication in the technological realm. On a cyber frontier, however, it makes sense that the natives-as-appropriated labor, those with the knowledge of the territory being explored and colonized, would be those with the knowledge of cyberspace. In this idealized dynamic, "those tech support guys" function, ideally, as native informants for the average person and as the "matter" for capital.

By promoting this dynamic, Kingston and other ads marketing the technologies can further encourage readers to associate computer technologies as the means to an economically empowered citizenship and the vehicle to an imagined frontier where we can stake a claim on such economic agency. With such power, an Avery advertisement boasts, we will have "more time to sit back, relax, and bask in [our] superhuman intelligence."[31] In these invocations of power and control, the function of technology becomes to make us more our "true" selves; that is, according to these ads, the subject without access to such technology is the victim of what Markley calls an "originary alienation" that can be "repaired" only by supplementing our imperfect selves with the latest technology. Those marketing

computer technology thus position the cyber frontier and our interaction with it as essential to our ability to occupy a mobile, empowered, futuristic, and always imaginary "coherent" subjectivity. In
fact, like the promotional material for Mars and the originary American frontier West, the rhetoric of many advertisements defines the
future in terms of the cyber frontier so that participation in cyber
colonial ventures seems essential for survival.

In this way, controlling an avatar is presumed to offer us the illusion of assuming what Kaja Silverman calls the position of "exem-
plary male subjectivity."[32] We are led to believe that we can occupy
an authoritative position, traditionally constructed as masculine,
which is aligned with the male "gaze" in much film theory. From
this superior position, which Katz ironically suggests is also an *equal*
position with other free godlike *individuals*, one has the "freedom"
to control cyberspatial environments without the constraints, laws,
rules, and regulations of the lesser, "real-world" American culture.
Katz and many of the ads I cite have promised us cyberspace as a
free democratic space and a free subjective space where we can be
ourselves without the very forces (whether psychological or ideological) that constitute our always incomplete selves. "They" have promised us contradictory things: cyberspace as a democratic space and
cyberspace as a realm of resources that "we" can control. In this
regard, cyberspace reinscribes the basic contradiction of the frontier: democratic, communal rhetoric versus self-aggrandizing indi-
vidualistic values.

Many advertisements have used the rhetoric of frontierism and
colonization as though the supposedly "powerful" selves "controlling" such technologies were not always and already being reproduced as citizen-consumers by virtual technologies. And, of course,
they are. The civic power that we are supposed to expect from our
various cyber experiences depends on our being able to have
greater mobility, control, and agency than those real-world citizens
who live, primarily, in the material world (the "nonfrontier"). Late-
twentieth-century ads for computers and computer programs conspicuously boasted that we need such technology to win the "war,"
or out-do others economically and socially, and support a frontierist
ideology of consumption predicated on our civic insecurities. Since
the compulsive, frontierist mining of cyberspace for value and civic
power produces fantasies of a powerful and coherent self only by
demonstrating that this frontierist fiction is a product of the very
technologies that the masculine agent supposedly masters, we must

recognize the ways in which these frontier technologies encrypt and inscribe narratives which do not answer questions of civic agency and citizenship but try to transcend them.

Notes

1. Frederick Jackson Turner, *The Frontier in American History* (1920; reprint, Toronto: Dover, 1996), 36.

2. See J. David Bolter and Richard Grusin, *Remediation: Understanding New Media* (Cambridge: MIT Press, 1998). Bolter and Grusin discuss the ways in which the prosthetic and transcendent functions of cyberspace stem from the same unrealizable desire to have unmediated access to the "real."

3. Jon Katz, "Birth of a Digital Nation," *Wired* 5 (April 1997): 186, 191.

4. By "postoriginary" I mean the period of American history which follows the official closing of the frontier in 1890 by the census bureau. The "originary frontier" would be, then, the American frontier West prior to 1890 when the frontier line still existed on census maps. See United States Census Office, *Statistics of the Population of the United States of the Eleventh Census: 1890.* Part 1: Progress of the Nation (Washington, D.C.: Government Printing Office, 1894).

5. William Gibson defines cyberspace as "a consensual hallucination experienced daily by billions." See William Gibson, *Neuromancer* (New York: Ace Books, 1984), 51.

6. While not the source of my formulation, the logic I identify in these ads is similar to the logic of the "supplement" which Derrida discusses in "Of Grammatology." See Jacques Derrida, *Of Grammatology,* trans. Gayatri Chakavorty Spivak (Baltimore, Md.: Johns Hopkins University Press, 1998).

7. Bolter and Grusin write: "Hypermedia and transparent media are opposite manifestations of the same desire: the desire to get past the limits of representations and to achieve the real . . . Transparent digital applications seek to get to the real by bravely denying the fact of mediation. Digital hypermedia seek the real by multiplying mediation so as to create a feeling of fullness, a satiety of experience, which can be taken as reality. Both these moves are strategies of remediation" (*Remediation,* 343).

8. Creative advertisement in *Wired* 5 (October 1997): 5, also, AST advertisement in ibid., 37.

9. Robert Markley, "Boundaries: Mathematics, Alienation, and the Metaphysics of Cyberspace," in *Virtual Realities and Their Discontents,* ed. Robert Markley (Baltimore, Md.: Johns Hopkins University Press, 1996), 504.

10. N. Katherine Hayles, "Virtual Bodies and Flickering Signifiers," *October* 66 (1993): 69–91.

11. Anne Balsamo, *Technologies of the Gendered Body: Reading Cyborg Women* (Durham, N.C.: Duke University Press, 1996), 127.

12. Nancy Leys Stepan, "Race and Gender: The Role of Analogy in Science," *Isis* 77 (1986): 261–77. In this article Stepan argues for what she calls the role of constitutive metaphors in science. In her view metaphors do not passively reflect a perceived scientific reality but constitute it.

13. That is, as Bruno Latour in *We Have Never Been Modern*, trans. Catherine Porter (Cambridge: Harvard University Press, 1993) and others suggest, recent cybercolonial narratives are neither fundamentally neoteric nor radical. They are, rather, very much like the imperial narratives of colonialization. Cyberspatial frontier narratives use the concept of frontier "freedom" to mask the colonial desire always imbricated therein. Such supposed freedom is dependent upon the desire to have a "masterful grasp" of the real, material world, a grasp that is—they suggest— inaccessible to us without technological intervention. However, it is, of course, also inaccessible to us through the cyberspatial frontier.

14. For example, in Chinua Achebe's novel *Things Fall Apart* (New York: Fawcett-Crest, 1990), British colonizers thought of themselves as missionaries who were superior to and more civilized than the Ibo people because they had "modern" guns and other technological instruments.

15. Michael Adas, *Machines as the Measure of Men: Science, Technology, and the Ideologies of Western Dominance* (Ithaca, N.Y.: Cornell University Press, 1989), 4.

16. Pioneer advertisement in *Wired* 5 (April 1997): 6.

17. One Web site uses the following text to introduce its material on outer space: "The legendary Captain of the USS Enterprise, James. T. Kirk, called Space the 'final frontier'. [*sic*] Although this may be true, the Net brings it that much closer to our home," <http://realguide.real.com/scitech/?s=finalfrontier>.

18. Compaq advertisement in *Sky*, February 1997.

19. Toshiba advertisement in *Delta Sky Magazine*, February 1997.

20. Micron advertisement in *Sky*, February 1997.

21. In his essay "The Emergence of an American Ethnic Pattern," Ronald Takaki defines the goal of the melting pot theory as "full assimilation to a new identity." See Ronald Takaki, ed., *From Different Shores: Perspectives on Race and Ethnicity in America* (New York: Oxford University Press, 1994), 21.

22. MCI advertisement on TV in 1996.

23. Katz, "Birth of a Digital Nation," 52, 50, 184.

24. This illusion of a choice is a key element in naturalizing the contradictions of liberal democracy.

25. David Tomas, "Old Rituals for New Space: Rites de Passage and William Gibson's Cultural Model for Cyberspace," in Michael Benedikt, ed., *First Steps in Cyberspace*, 5th ed. (Cambridge, Mass. MIT Press, 1991).

26. See Edward Said, *Culture and Imperialism* (New York: Vintage Books, 1993).

27. GeoCities advertisement in *Glamour*, February 1996.

28. Webtv advertisement in *Self*, December 1996.

29. Kingston Technology advertisement in *Sky*, February 1997.

30. Gayatri Spivak, *In Other Worlds: Essays in Cultural Politics* (New York: Routledge, 1987), writes that zero-work is "work not only outside of wage-work, but, in one way or another, outside of the definitive modes of production" (84). Along the same lines, Robert Blauner writes in "Colonized and Immigrant Minorities" that " 'New World' events established the pattern for labor practices in the colonial regimes of Asia, Africa, and Oceana . . . The key equation was the association of free labor with people of white European stock and the association of unfree labor with non-Western people of color." In *From Different Shores*, ed. Takaki (152).

31. Avery advertisement in *Sky*, February 1997.

32. Kaja Silverman, "Dis-Embodying the Female Voice," in *Re-Vision: Essays in Feminist Film Criticism*, ed. Mary Ann Doane, Patricia Mellencamp, and Linda Williams (Los Angeles: University Publications of America, 1984), 134.

Hegelian Buddhist Hypertextual Media Inhabitation

David Kolb
Bates College

Our moment has absorbed the linguistic turn of modern epistemology, to move now into a pictorial turn . . . The challenge to the disciplines of Arts and Letters is to invent or design the practice of this syncretic writing . . . The basic reality of the pictorial turn is that the site of invention of the next stage in the evolution of writing is taking place within the institution of entertainment . . . Unfortunately the media literacy movement still formulates this moment almost exclusively in terms of literacy, wanting to make citizens more critical of what they consume in the media.[1]

W E know the gamers. The teenager hunched over the controller jumps, weaves, and kicks Super Mario through the palace and on to a new level, or blasts his way through Doom or Quake, or lingers silently with *Myst*'s moods and puzzles. He's immersed in the game.

We know the critics. *The New York Times*. Thumbs up or down. Academics writing about novels. Plato banishing the poets. The critic stands apart from the game and issues judgments.

The critics condemn the gamers. Get a life. Get some art. But in the media age will criticism (or art) survive?

Art has been under pressure from economic and cultural normalizers throughout its history. In this century, it has been attacked by artists themselves. Duchamp, Warhol, the avant-gardes and neo-avant-gardes have been joined by philosophers and social theorists in questioning or transgressing the categories of high art. The institution of criticism has scarcely fared better at their hands, even as they stake out critical positions.

Art and criticism continue. Art institutions are going strong and are ever more efficiently administered—not an altogether happy situation. In the culture industry we distinguish dinnerware from ceramic art, illustration from painting, and soon mundane from artistic virtual worlds. Such distinctions add market value and class identification. They may even have intellectual content. But art has difficulties even wanting to maintain distinctions in the media age.

As for criticism, it is often reduced to either information or showbiz. In the clamor of the media, do we need critical standards to decide where to spend our time? We certainly need information, since attention and time are scarce resources to be distributed wisely. Where once the tragedy festival, the cathedral, or the nobility's display provided a focus for attention, we have an oversupply of proffered foci. Our communities are not so local or so homogeneous as before. Which cathedral or opera or painting or cultural event will we involve ourselves with this week? There is no one center that gathers "us" and "our" art. So we have lists that tell what events, objects, experiences—what communities—we can expose ourselves to. Beyond *TV Guide* waits *The New Yorker*, which lists and judges, and then *The Nation* and *The New Criterion* and many others that only judge. Such voices often distinguish between proffered experiences that fit smoothly into an audience's values and expectations and those that oppose or challenge them. That may be taken as a reason to avoid or to embrace the experiences.

I wonder about criticism directed at immersive cultural artifacts. This awkward term is meant to gather those creations that open an explorable sensory context. Immersive artifacts encourage us to ignore "outside" stimuli while we explore an offered environment. This might be a computer game, or a virtual world, or a MUD, or a chat environment with a distinctive graphic atmosphere and self avatars. We are beginning to build such artifacts, and the science fiction dream of lifelike immersion will eventually come true.

> Tonight, like every night for the past eight months, tens of thousands of players will log on to Brittania, a fictional online universe. They'll come to embroider upon make believe lives as healers, fighters, mages, and rogues. And they'll stay—up to four hours each—because of the seductive quality of pure immersion . . . On some nights, more than 14,000 players are logged on at once. More than half of them log on every day . . . The towns, forests, and dungeons of Brittania are more than just intricately rendered; details are meaningful—you can pick up and read a book on the library shelf or play a game of checkers in the tavern . . .

Britannia occupies some 32,000 screens, with 15 major cities, 9 shrines, 7 dungeons, and vast stretches of uncharted wilderness. As more and more players put down roots, the landscape . . . changes accordingly.[2]

While such computer-aided environments are the most obvious case, there are many other immersive artifacts. A less interactive immersion is already available with films and TV, the latter perhaps offering linked constellations of miniworlds prefigured in the alliances among cable channels. But high tech is not the only way to create immersion. Live action role-playing games such as Dungeons and Dragons, or Assassin and other extended games played by college students, use everyday objects and spaces and their players' imaginations to immerse their participants in jointly maintained fictions. Series of books and TV programs may create a world that is returned to and incites further exploration. In the media rush, some immersive artifacts become brand names: Star Trek, Star Wars, or Disney stories. These sell repetitions of themselves. But an immersive artifact is more than a brand name; it offers a world with room to move about. Technology can enable that exploration in real time, often in the company of other participants.[3]

Such worlds assert themselves as relative totalities; yet they are part of our temporal experience of many worlds, and part of the net that is coming into being. Whether we think in terms of a walled-off world or a linked net of worlds they cause problems for criticism.[4]

From Plato's attack on the Sophists down to postmodern complaints about consumer culture, critics have worried about the power of rhetoric and image to mold beliefs and values while suppressing critical examination. Immersive artifacts may manipulate people even more thoroughly, shutting out the critical voice and keeping their inhabitants busy with no time to think. To combat this danger, where should the critic stand, and how will the critical voice be heard? Or is the critic's only choice to stand and speak?

Traditional criticism locates itself at a distance, immune to manipulation because based in clear principles derived outside the images or artifacts being criticized. Often the critic seeks to disengage from the object a set of propositions that can be attacked with the tools of logic and argument. But criticism of art and imagery that reduces them to implicit arguments or networks of beliefs has never been very successful. Try extracting the content of Shakespeare's *Hamlet* or Michelangelo's *David* into argumentative form. Nor is such criticism adequate for today. The media influence us on many levels only some of which are amenable to argumentative treatment.

Yet in our age of imagery run riot, there has been a reversal that would have surprised Plato and Socrates. Rather than remaining the object of distanced criticism, art in this century has itself developed explicit strategies for questioning attitudes and cultural power. Often these artistic strategies involve the collaged juxtaposition of abruptly discontinuous fragments of imagery or belief systems, violating the borders of cultural spheres. Both the heightening of boundaries and the erasure of boundaries become critical tools within art. Other recent art highlights the multiplicity behind apparent unities. The critical strategies I suggest in the text resemble these artistic moves.

Distanced critical analysis does not work so well with immersive artifacts. Traditional criticism does not easily reach their participants. This might not seem important to people used to consulting authorities located in separated critical forums, but it is not enough to have distanced analysis going on in its own enclave for its own specialized audience. How do participants-consumers of immersive artifacts become more self-aware and make more nuanced judgments than turning off the tube or modem? This will not happen by increasing the readership of separated critical forums.

Nor is it likely to happen by inserting traditional critics into the nets or the immersive worlds. If you want to do critical discussion where the people are—criticism of television on television, or of hypertext in hypertext, or of a virtual world in the virtual world— you find that immersive artifacts surround distance and use it for effect. Traditional criticism frames itself as distinct from its framed object of analysis. But frames have become items within the flow rather than borders around the fray.

The critic becomes part of the show. The expert's arguments become something to be enjoyed—the more passionate the better. The spectators view the game, but their allegiances need not be questioned. Standard critical stances and tools are co-opted. In an age of link buttons and of Webs without edges even separated critical forums can become one more channel in the media show.[5]

This is an extreme version of a perennial problem. Socrates tried and failed to make people distinguish him from the Sophists; he saw himself as a critic of performances; they saw him as another dangerous performer.[6] Immersive cultural artifacts make this situation worse because they segment the common place of public meeting and discussion, fragmenting the agora where Socrates once could encounter any citizen. Critical efforts to establish an authoritative meta-agora above the fragmentation produce just one more place to visit.

So, where does the critic stand? Does the critic stand? Or move? The critic can get lost in the funhouse, immersed in the artifact and playing by its rules, or the critic can wander lost on the Web, or the critic can stay distanced wielding outside principles and norms. These options do not seem quite right. Can the immersed inhabitant truly judge? Can the wandering critic be heard? Can the outside critic really know what is being judged? A common solution is the anthropologist participant-observer-critic who enters an artifact or joins the net armed with insights and principles from an outside framework. This critic then distributes judgments. Such outside judgment remains important for many purposes, but I am seeking other modes of critical interaction.

Traditional critics bring to their task principles formulated outside the language of the work being criticized. Such principles might concern the nature of art or of culture and discourse. They might concern goals to which our creations should be subordinated. They might be principles of form. To apply such principles to a cultural artifact, the critic reformulates or redescribes the artifact in terms that will connect it with the principles. Formalist critics redescribe the object in relation to ideal standards. Other critics might reformulate the artifact in a narrative about earlier masterpieces, or about ongoing themes, or about class struggle or gender domination. One might judge that a video game reinforces sexual stereotypes or an immersive world pushes the values of consumption. But I want to suggest other critical modes that can operate inside the immersive artifacts and on the nets. These other modes do not stand and pass judgment.[7]

There are critical modes of inhabiting that do not reformulate the language of the artifact. They stay within its language and rules and find there spaces for critical interventions. This is possible because these modes do not accept a presupposition which lies behind both the fear of uncritical immersion and the desire to establish critical distance. Both presuppose *the control of meaning*. In this phrase the "of" should be understood both as the critic's ability to frame a stable meaning to be studied and the meaning's ability to control a world's inhabitants by surrounding them in a seamless whole. Both presume that meanings can be woven into a single tight simply located unity.[8] If we question this unity or its stability, then inhabitation cannot be a simple submission; it will have its distancings and porosities. Modes of criticism can live within these transitions and distensions rather than on secure metaplatforms.[9]

One such critical inhabitation is familiar enough. The image of the teenage boy immersed in a video game gives way to the image of the teenage hacker finding ways to manipulate a computer game. But he doesn't need to be a professed hacker; there are books available that give him solutions to the *Myst* puzzles. There are utility programs that let him enjoy the thrill of the game and also beat or change its rules. MUD wizards embody this doubled inhabitation. They live in the MUD as both its participants and as its software engineers. This doubled role is a way of being inside the artifact. Even when they are involved with issues of code and scripting, they are not so much alternating being inside and outside the game as they are inside in a way such that the two aspects of that inhabitation feed off each other: programming the MUD by itself would not be so fascinating if one did not also have a role in the virtual world, and the role becomes more vivid as one gains power over the world's infrastructure.

Even passive media such as TV can stimulate such doubling: *Soap Opera Digest* offers its readers both a deeper immersion in the plots and a sense of being behind the scenes examining the production process. The passive media can open very actively shared areas: Star Trek shows and movies give birth to fan-authored magazines and get-togethers that offer further explorations of the shared world, looks behind-the-scenes, and a chance to create one's own story variations.[10]

This kind of manipulative involvement allows you to criticize an immersive artifact within its own parameters. You need not redescribe the artifact's rules and qualities in some distanced critical discourse. Having power within the world you can make it be more what it already wants to be. Or you can bend it gradually toward what you want it to be. The will to power is strong in this mode of inhabitation. Control and self-affirmation are prized, with or without social interaction. The standard image of the teenage hacker is of a loner, but this kind of activity can also be intensely cooperative. Whether alone or social, this is not simple immersion, since the doubled roles provide built-in room for discussion about the world as you inhabit it.

While the hacker can criticize and change arbitrarily large elements, hacker inhabitation does not easily challenge the overall teleology of the immersive artifact. But this might happen if the hacker creativity were joined with a sensitivity to the overall tenor of the immersive experience, the feeling of its life, and the powers and relations it assembles. Does participation in the artifact increase or weaken one's being? Such Deleuzean questions could direct activities as the hacker morphs into the artist, gathering experiences and adding to the assemblage of events and singularities.

Such immanent criticism can give some body to the hacker role. Deleuze's rhetoric of "lines of flight" and "nomads" suggests fleeing rather than remaking the artifact.[11] But in the case of digital artifacts the way they are transmitted and constantly recopied, reentered, and linked means that there is less difference between reforming and creating. One linked world can be a part of, and an addition to, and a criticism of, another world.

Besides the hacker, there are other modes of immanent criticism that do not necessarily intervene on the infrastructure of the immersive artifact. They motivate different activities within the artifact, or different relations across nets of artifacts, and they change one's relations in the artifact's world. Philosophically their presuppositions conflict with one another and with the Deleuzian approach.[12] It is not my purpose to settle such disputes here, but only to show that immanent criticism is a real possibility.

Immersive artifacts involve internal motions across transitions, links, and differentiated contents. Imagine a strongly temporalized inhabitation that lets those moves happen and pays minute attention to them, yet also lets their borders and connections and flows be as in-betweens but not as fixed oppositions. The attempts of one content or virtuality to define itself as total or primary or separate are taken as just that, attempts, to be noted rather than accepted or rejected. Interaction, infection, interpenetration among the contents or stages are not repressed for the sake of some tightly held identity, nor insisted upon to attack identity. Imagine a letting-be that refuses to be drawn along but does not hold back, that allows languages and movements to be themselves but, because it does not have to be identified with any particular result, is not pulled about by the unfoldings and not seized or divided in the conflicts. It is an alert seeing where things go and where they end, what feelings and experiences they create, and where the content clutches and clings and opposes and tries to be more than it can be. Such an inhabitation does not rest on any particular content or principle as a base. It is less vulnerable to manipulation and distraction because it does not grasp at any bait but only observes it as such. It is made possible by the spacing inherent in temporal experience and the open texture of meanings. Because there is no clinging to current rules and definitions this mode of inhabitation can accept influences across borders and echoes from other worlds. The artifact is not reformulated, but one refuses to be stopped at borders or to cling to official purposes.

The Buddhists have a word for this attitude: nonattachment, neither clutching nor rejecting. They have a word for the process: mindfulness. They have a word for its effect: compassion. And a word for what it leads to: skillful means.[13]

This attentive self may not seem critical enough, for it does not stand off and judge. But this mode of inhabitation criticizes, by noticing them, attempts to grasp or to divide or to force integration; it refuses to be drawn into desires for totality. It offers resistance without aversion.

Nor need nonattachment be purely contemplative. If intervention happens, it could be like the Buddha's actions told in the Jataka stories. For instance: in an earlier lifetime two future Buddhas, Maitreya and Gautama, encountered a starving tigress with her suffering cubs. Rather than fleeing the snarling beast, Maitreya set off to find food for her. This stretches the rules of the game for dealing with wild animals. But then, when he returned, Maitreya discovered that Gautama had taken a still more unusual and decisive action: he had fed himself to the hungry tigress. Such an action is outrageous according to the norms of the everyday world, and so challenges our motives and norms. The bodhisattva's action resembles that of the poet making a new metaphor and changing language. It makes new voices audible and opens up new shapes of life. This is a different spirit than the will to power of the standard MUD wizard. It might open up the social practices in shared worlds, making people aware of new possibilities and new ways of being. This might question the normative frames within which people are acting, or repurpose their activities and change the stakes.[14]

There are other modes of immanent criticism that are more oriented to the detailed structure of meanings within an artifact. One such mode could be derived from Hegel. In the introduction to his *Phenomenology of Spirit* Hegel raises the age-old problem of the criterion: how can we find ways to criticize our self-conceptions and cultural productions when the criteria that we invoke are themselves cultural productions in need of justification by still other criteria? Hegel avoids the threatened regress by refusing the demand for foundational criteria. He suggests a process of just looking. Letting a self-conception or a cultural artifact be itself can be a way of letting its internal tensions and contradictions express themselves. We could raise questions about Hegel's theory and the necessity of the sequence of self-conceptions he presents, but his mode of criticism suggests a way to inhabit immersive cultural artifacts. Hegel would say: you don't need to import criteria from outside, because the artifact already has a built-in self-understanding and goal. Let this show its dialectical transformations and loss of self-certainty.[15]

Might there be a kind of self-criticism that happens when the artifact tries to be itself? The point is not that the artifact aims at some effect but falls short. Rather its goal is part of a self-conception that covertly depends on other relations not yet included within the self-conception. Falsely absolutizing its situation, a mode of living does not see its constitutive dependencies and interrelations. When it tries to realize itself in this narrowly conceived way and world, it fails and develops an enlarged self-conception. Self-conceptions and cultural forms can reveal their connections even when they do not know them.

Hegel is more concerned with overall modes of being in the world than with the details of individual cultural artifacts. He studies the ways that cultures, religions, or artistic genres develop out of one another. But this is not wholly distinct from studying the shapes taken by shared inhabitation of virtual worlds, even those as impoverished as current computer games.

Imagine then a mode of letting-be within a space created by the self-conception and grammar and values implied in the artifact. It makes that space explicit and reveals inbuilt tensions and motions. It is alert to how the possibilities change as the self-comprehension tries to realize itself. Inhabiting an immersive artifact in this way would be to stay within the artifact's terms as they change, adding only a memory and a description of the transitions. This creates a narrative in which the current self-understanding is only one phase.

The classic overnarrative of this sort is Hegel's story about how the necessary structure of our being is the coming to self-consciousness of that necessary structure itself. But Hegel is not alone. There are more recent modes descended from Hegel that also remain with the language or system of the inhabited artifact but refuse to take that language or system as fixed or final.

One such mode of inhabitation would act in the artifact's world somewhat in the manner in which Derrida reads a text: attentive to slippages and to covert dependencies on devalued and supplementary elements. Seek the ways in which the world's unities come about as effects in a field that these unities do not dominate even though they announce themselves as doing so. The deconstructive inhabitant looks to perform gestures that put stress on the standard rules and divisions and make them show their slippages and covert dependencies. She might reuse and recombine bits of the world in unexpected ways, making unconventional moves or links that seem inappropriate but reveal hidden connections or put pressure on invisible walls.

This mode might lead to a local equivalent of what hypertext theorists have called "writing all over the interface." Items that are part of the machinery of an artifact's world (menus, lists, maps, configuration files, margins, indices, and so on) can be made part of its poetry. This brings the inhabitant up short against machinery that refuses to stay in its subordinate role, and opens it to new possibilities. For instance, in Michael Joyce's hypertext short story *Woe*, the names of the links on the maps are arranged so that they read as poems. Some of the clues in John McDaid's portrait-adventure-detective-story-world, *Uncle Buddy's Phantom Funhouse*, are to be found by viewing the hypercard coding. In the title essay of his collection *Writing at the Edge*, George Landow collects many transgressive and playful reuses of what might have been only background machinery.[16] In artifacts such as MUDs where the infrastructure is available, such interventions could help question and undercut the naturalness of wholes and transitions that get taken for granted even in such completely artificial environments.[17]

I have suggested modes of inhabitation that do not reformulate immersive artifacts into a foreign critical language. Yet in their different ways they refuse to take asserted unities and structures as final. They sense borders without getting caught in fixed oppositions. They let meanings develop and cross and criticize their prior selves. They take unities and closures and oppositions as effects within a relational network rather than as given poles and borders. They react to the overreaching built into such effects. If closure is an effect rather than a given, these modes of criticism let this be shown within the net or the artifact's own terms, rather than reformulating the artifact and subsuming it within another closure set up by exterior principles or narratives.

There are philosophical battles to be fought over the relative priorities and the possible reductions of one mode to another. I am taking no sides in such disputes here; my point is that these offer less distanced ways to have critical effects. Deciding which modes are more fundamental than others is an important task, but such a decision would not reduce them all to a single mode.

But what kind of effects might these have on the inhabitant-user or on the immersive artifact?[18] The kinds of critical interventions depend on the degree of interactivity possible in an environment. The classic video game Pacman has one control, for the direction of movement of the player's representative on the screen. Speed and activity are not variable. Nintendo and Sega game units have controls that can vary the direction and timing of several activities. Games that allow verbal input bring more possibilities of control and intervention, as with *Adventure* and its descendants, but they can only parse simple commands. This changes when the computer becomes a medium of interpersonal interaction. MUDs and MOOs allow indefinitely many kinds of interventions. As technology for virtual reality matures there will be a similar range from passive VR rides to mutually created and modifiable virtual worlds.

Because of their relatively free sociality, MUDs and their future VR parallels offer the widest possibility for critical inhabitation. The active attention suggested above could sensitize users to dependencies and rigidities built into their world. This could change the inhabitants' stance within the world, the way they relate to other inhabitants, and eventually the rules and features of the environment itself. Deviant behavior could make points about accepted norms. Rooms could be built that open new possibilities. Conventions could be challenged verbally or by other consciousness-raising maneuvers. MOO software often contains explicit tools for enhancing debate about the structure and rules of the MOO itself.[19]

A jointly composed hypertext Web offers analogous possibilities. For instance, I might criticize your contributions to the text not by direct argument against them, but by linking them in unexpected ways or by reusing them in an unexpected context.

When there are fewer modes of input within the artifact, or its infrastructure is not available except to the hacker, there are still critical possibilities. It may be possible to break the rules and create new behaviors or to stimulate a metagame within the game. In a shared game there still might be occasions for bodhisattvalike or deconstructive moves that go against the teleology of the game and open up unexpected modes of sociality. Remember Gautama and the tigress. Would one really try laying down one's arms in a combat game like Doom? This would make little sense in the solo version, because the computer cannot acknowledge such a move, but even such solos can prompt reflection.[20] In a social version that allowed enough flexibility in responses, refusing to fight might challenge the players to interact in new ways. Espen Arseth points out that shared interactive worlds are not best conceived as "games," since "any system that must regulate its discourse by social pressure and convention rather than by clearly defined regulations is more than a game—both more real and more perilous."[21]

In cases where the artifact's structure and goals are fixed, such as *Myst* and other solo CD-ROM adventures, only the hacker mode of inhabitation can change the immersive world itself. But the other modes I discussed can change the attention of the user, who would be more aware of the structure of the world and the interplay among its parts and values. These modes emphasize how the temporality of the experience allows but overflows borders, including the borders of the game.

Solo computer games may seem far from the world of high art, but the player's concentration on the game resembles the quasi-religious attitude enjoined on us in front of approved masterpieces. Just as recent theory undermines that passive reception of art, so it could affect the consumption of immersive artifacts.

Even games that are quite limited may still offer opportunities for changes in the attitude or stance of the players. In *SimCity* the underlying algorithms are chaotic, so there is no one state that wins the game—the aim is to plan and administer a survivable city. This could lead to discussion among players about what makes a city livable. In solo video arcade games one has only limited interactions against implacable opposition. But it is still possible to change one's own mode of inhabitation, which is never as simple as it might seem: consider how people play different arcade games one after another, making comparisons and judging the current state of the art.

Similar linkage effects also occur where inhabitation may appear totally passive, as in channel and Web surfing.[22] Even the most passive immersive artifacts, such as theme park VR rides, still offer the possibility of different awareness during the experience. That awareness need not be a constant comparison with what is outside, but could be awareness of interior relations and spacings and connections, as described above. Such more critical active attention could affect the evolving experience.

But can changes of attention really have potentially critical effects? Several objections will help clarify the issues. The first objection is that immersive media are experienced in a state of distraction. Not because we are doing something else (though some media can be used as wallpaper) but because the media are immersive, that is, they provide an environment that is richer than any single focus. The whole experience could be in a state of distraction. Think of a child in an amusement park running from thrill to thrill without ever fully attending to any one event because each is so infected with the lure of the next and with the contentless promise of even greater to come. TV and Web surfing can bring this distraction either with the child's eagerness or with a desultory boredom. Such distraction increases our chances of being manipulated, since it reduces awareness of meanings or connections other than those now being fed to us. This shows how immersive technology can inhibit the attention needed by critical inhabitation.

However, the crucial issue is not distraction but forgetting. Even a distracted state can be mindfully experienced as such in the Buddhist mode, or inhabited in its transitions and demands. The real enemy of critical inhabitation is a sequence that drops its past as it goes. Although the modes of inhabitation I have suggested do not demand a metastance outside the artifact, they do require resolute attention to the movement and the temporality of experience. They fight forgetting and quick thrills.

Hacking gamers or MUD wizards slow down and work on the rules and infrastructure, then speed up to slalom through part of the world. There is Buddhist attention to qualities and temporal passage. There can be attention to the quality of the experience, even to its speed and distraction, if that is what is going on, as well as to the links and associations and carry-overs across the sequence. Hegelian and deconstructive modes demand more active remembering. Their attention is not as wide as in the Buddhist mode but is more focused on just those connections and transformations ignored in the quick thrill.

A second objection is almost the reverse of the first. If distraction leads to underattention, there can also be overattention. Immersive artifacts can get us overfocused on specific problems and goals, with no slowing down to smell the virtual roses or consider contextual structures and associations. Or the immediacy of some desired thrill or sexuality can narrow our focus to the present gratification, forgetting past developments and future consequences.

There is no avoiding the need for awareness on more dimensions than the immediate task or thrill. But it is possible to focus on the task or thrill without losing sight of its context and presuppositions, though this takes a more complex temporal rhythm. Criticism takes effort, though in this case it is not the effort of constructing a separate critical discourse. It takes advantage of the internal differentiation built into our temporal experience of meanings and activities.

A third objection comes from a different direction but has the same answer. We can be caught—the remote control and the mouse both liberate and enslave. They bring flexibility yet predefine our reach. They empower the subject to say "no" to this or that presentation even as they pin the subject within their particular network of presentations. The user can turn off the machine, but while it is connected the administrators of content will try to keep the user within their set of linked channels or sites. A cow, or a mind, can be controlled by giving it a big enough field to wander in. The fences remain.

This objection would be more powerful if we were indeed caught within a single artifact. But if the remote control and the mouse pin us within a labyrinth, it already contains multiple worlds and artifacts. That multiplicity and its border crossings can challenge preplanned meanings and excitations. Even if there were no choice at all, that would still not forbid critical attention. John Cage composed several pieces of music in which radio stations were tuned in and out on a fixed schedule. Listeners heard segments of whatever happened to be being broadcast on those stations.[23] Cage created attention and connection. You might call his creation a kind of controlled surfing, but it encouraged a complex attention. The individual segments were heard three ways at once: as particular sounds with their timbres and qualities, as recognizable types of programs, and as segments in Cage's work that played off one another and generated meanings across the discontinuities within the piece.

If one were to channel or Web surf with such dimensions of attention, new meanings would be created by the juxtapositions even if one could not choose the elements in the series. There is no control of meaning, so even channel surfing can generate new meaning across its conflicts and contrasts. What is needed is attention within the sequence itself, rather than instant forgetting in the search for thrills.

What I have been doing, in effect, is trying to move the conception of immersive media toward the kind of attentive literacy sought for the links and crossings of hypertext. Hypertext offers more than a sequence of on-off commands with the remote. There is always linking, mixture, and memory. These can be experienced attentively without being formulated in another language.[24]

Further technological advances will allow active linking rather than passive reception, with individual or group creation of new content. This is already beginning on the Web with those millions of homepages. Imagine them becoming homeworlds. Linking would become even more a mode of expression and creativity. We might have something resembling what Vannevar Bush prophesied: worlds that were trails of references, collections, critical comparisons—but all of these involving much more than just the information that was Bush's concern.[25] Imagine the construction of immersive collages, virtual space conceptual art installations including one another, new meta-artworks, super-Rauschenbergs using duplicative and linking technology. These would benefit from the kind of critical inhabitations I have been discussing.

With the technology available in the past hundred years, art developed the self-critical strategies already mentioned for examining its own conventions, institutions, and context. Those techniques of self-reference, collaged juxtaposition, disconnection and impertinent connection, ironic reuse could well become tools of self-expression and self-criticism on a more mundane level. Advertisements are already re-forming our sensitivities in that direction.

You might object that it is more likely that all the links will be premade corporate products. I think not. But even in that worst case remember John Cage. There would still be no total control by meaning or control of the generation of meaning. There will always be temporal sequence and its mixture and border crossings. There can still be critical attention to meaning beyond what is intended by the makers. So there will always be room at least for self-criticism of the user and possibly for unexpected interventions against the grain.

The modes I have described do not do away with the need for distanced argumentative criticism. Political and cultural criticism may need to back off from immersive artifacts and reformulate them within larger narratives and critical vocabularies. But that is not the only way. Meaning cannot be controlled; it opens new possibilities inside the immersive artifact. I have suggested modes that invoke this openness of meaning and the temporality of our experience. They allow the creativity of impertinent moves. The critical sensitivities involve more than sets of propositions and their inferential connections. We will be impoverished if we envision all criticism on the model of a logical argument or on the model of a judicial examination.

I began with two images, the gamer and the critic. Here is another: the writer creates a hypermedia novel or cyberdrama, hoping the readers will become immersed, but it is also an active hypertext, so readers are expected to add to the Web. The critic enters the Web and loses her metaposition, but she joins everyone in the possibility of reflection and the movements of meaning.[26]

Notes

I would like to thank Mark Bernstein, Charles Ess, Simon Buckingham Shum, and Gregory Ulmer for their insightful comments on an earlier version of this essay, and I thank Stuart Moulthrop and Nancy Kaplan for the inspiration provided by their essay "They Became What They Beheld: The Futility of Resistance in the Space of Electronic Writing," referenced in note 26.

1. Gregory Ulmer, "A Response to *Twelve Blue* by Michael Joyce," *Postmodern Culture* (autumn 1997), secs. 13–14. Ulmer raises issues related to those I discuss here in his "The Object of Post-Criticism," in *The Anti-Aesthetic: Essays on Postmodern Culture* (Port Townsend, Wash.: Bay Press, 1983).

2. Amy Jo Kim, "Killers Have More Fun," *Wired* 6 (May 1998): 141–43. Kim goes on to discuss problems with the culture that has evolved in the immersive environment, *Ultima Online*.

3. Janet H. Murray's *Hamlet on the Holodeck* (New York: Free Press, 1997) offers a useful survey of current and possible future immersive artifacts (chaps. 2 and 9) together with remarks on the psychology of immersion (chap. 4).

4. The open net and the enclosed world seem opposed, but enclosure is an effect created in the net. There will be connections even when there are no explicit links, as happens now with search engines.

5. A critic appearing on a TV program may establish her celebrity more than her critical authority. While an audience may come to trust some celebrities as sources of beliefs and values, this does not follow immediately from the intensity or the location of their performance.

6. Socrates wanted centered people to stand aside from the civic fray and be open to argumentative questioning. The Sophists wanted to control the civic fray by manipulating its uncentered multiplicity of desires. In the rhetorical melee, Socrates' attempt to establish a distanced critical dialogue was seen as just one more partisan cabal.

7. The modes of critical inhabitation outlined below do involve judgment in the sense that their attitudes and moves could be formulated in propositions that would involve evaluations. But the critical moves do not consist in the power-full enunciation of those propositions. Nor are those propositions based on the critic's immersion in some other world taken as a critical base.

8. The phrase "simply located" derives from Alfred North Whitehead's discussion of the "fallacy of simple location" in *Adventures of Ideas* (New York: Macmillan,

1933), 157–58. Whitehead criticizes philosophies that take qualities and things as firmly located in one separate location in a spatio-temporal whole composed of such immediate next-to's. However one evaluates Whitehead's ideas regarding physical space, I want to apply them analogously to meaning: there is no simple location in logical space.

9. This is relevant to worries about the possible totalitarian effects of immersive artifacts. Such experiences can weaken critical and emotional distance, opening the user to manipulation. I want to argue, though, that metadiscussion and distance are not the only critical tools. The impossibility of complete closure or control of meaning opens other modes of critical inhabitation. The issues of memory and attention raised later in the text are important in this regard. In addition, no matter how immersive they may be, immersive artifacts occupy only a portion of one's time. Perhaps a greater worry should be the monotone experience created when a subcommunity tries to color all accesses—newspapers, radio stations, music, magazines, books, films, as well as games—all with a party line. These are situations where criticism by reformulation in a larger framework may be very helpful, but they are also open to the kinds of critical interventions discussed in this essay.

10. Fanzines create family variants of the "official" shared worlds. See the discussion of fan culture in Murray, *Hamlet on the Holodeck*, 41, 85, and Gregory Benford's remarks in the "Afterword" to his *Foundation's Fear* (New York: Harper, 1997) about interaction and improvisation in genre literature (420). Recently some of the brand-name worlds, such as Star Trek, have threatened legal action against fan publications and fan Web sites that produce their own variant stories and images. This stems from worries about the control of intellectual property and fears of stories at odds with the standard characterization, for instance stories about a sexual relation between Kirk and Spock. Such exploration/variation seems inevitable with shared worlds. As a publishing phenomenon it is at least as old as Cervantes varying the old romances to produce *Don Quixote*, which was followed by other authors writing unofficial continuations of Quixote's adventures, which were followed in turn by Cervantes trumping them all by writing the second part of *Don Quixote*, where the Don meets some characters from the unofficial continuations, who then admit that he, rather than the gentleman they met in their books, is the true Don. Today's reproductive technology makes such play inevitable.

11. For examples of this widespread rhetoric, see the section "Nomadology," in *A Thousand Plateaus* (Minneapolis: University of Minnesota Press, 1987).

12. Deleuze claims that not all definiteness comes from meaning structure and relation; this disagrees with the presuppositions of the Hegelian and perhaps the Buddhist modes discussed in the text. Deleuze's terminology suggests a campaign against structure, but he also says that "staying stratified, organized, signified, subjected—is not the worst that can happen; the worst that can happen is if you throw the strata into demented or suicidal collapse, which brings them back down on us heavier than ever. This is how it should be done: lodge yourself on a stratum, experiment with the opportunities it offers, find an advantageous place on it, find potential movements of deterritorialization, possible lines of flight, experience them, produce flow conjunctions here and there, try out continuums of intensities segment by segment, have a small plot of new land at all times." In Gilles Deleuze, *Essays Critical and Clinical* (Minneapolis: University of Minnesota Press, 1997), 160f. This quotation raises issues about the place of memory in Deleuze's theory.

13. One might conceptualize nonattachment in immersive artifacts using the somewhat different notions found in the Hindu *Gita* or in Taoism. But I find most useful the links to the ideas of compassion and skillful means in the bodhisattva's activity. This Buddhist mode has many resemblances to the Deleuzean mode mentioned earlier, despite Deleuze's claim that Buddhism is a polar opposite to his own Nietzschean affirmation of life. See *Essays Critical and Clinical,* 133.

14. Jataka stories draw moral lessons from episodes in the previous lives of the Buddha. They depend on the cosmology of reincarnation, and in that context there are understandable motivations for the surprising actions described in the stories. But the stories can also be read independently of any foundational cosmology, as challenges to ordinary mores and motives, in the way that a surprising new art form or new kind of political association can change our landscape of possibilities. The underlying philosophical issue here is whether the surprising move must always stem from some already formed narrative or principle. It seems to me that the surprising move is often itself the discovery of such a new direction of meaning, but there can also be moves whose effect is to loosen or challenge meanings and mores without providing a formed alternative.

15. G. W. F. Hegel, *Phenomenology of Spirit,* trans. A. V. Miller (Oxford: Oxford University Press, 1977), §§81–85.

16. Michael Joyce, *Woe,* in *Writing on the Edge* 2, no. 2 (spring 1991), on a disc included with the journal; John McDaid, *Uncle Buddy's Phantom Funhouse* (Watertown, Mass.: Eastgate Systems, 1992), on disc with audio tape and printed material; George Landow, ed., *Writing at the Edge* (Watertown, Mass.: Eastgate Systems, 1995), on disc.

17. Although I don't develop them further here, there are other possible immanent critical inhabitations. One could be developed out of the "carrying forward" described in the writings of Eugene Gendlin. See the discussions in *Language Beyond Postmodernism: Saying and Thinking in Gendlin's Philosophy,* ed. David Michael Levin (Evanston, Ill.: Northwestern University Press, 1997). Another might be formed around Paul Ricoeur's ideas about metaphor and narrative as refiguring experience and opening new spaces of possibility: see *The Rule of Metaphor* (Toronto: University of Toronto Press, 1977) and *Time and Narrative* (Chicago: University of Chicago Press, 1984–88). These might be called "hermeneutic" but not with the sense of an investigation searching to uncover unified hidden meaning. They would attend to the temporal carrying forward and change of meanings, aware of the fragility of these transitions and open to ways in which the present results from previous transitions. Both these and the deconstructive approaches deny the forced choice of immersion versus distance, as well as the presupposition that meanings can be controlled. Both urge attention to the stretched-out happening of meaningful artifacts.

18. Do these modes of criticism still depend on metapositions outside the artifacts? There is a sense in which they depend on very formal notions about the conditions of possibility for meaning and for closure effects. But such notions are not involved as premises supporting pronouncements of judgment.

19. Even when such tools are not available, people may still find ways to demand change. "A few weeks after *Ultima Online*'s release, a player named Mohdri Dragon initiated one of the game's first public displays of civil disobedience, to call attention to Origin's lax response to numerous unfixed bugs while it built new features.

Hundreds of players gathered together in the capital, stripped their characters naked, and stormed the castle of Lord British—a k a Richard Garriott, the real-life creator of the *Ultima* series. Once inside the castle, the players drank themselves silly, trashed Lord British's throne room, and protested loudly, much to the amusement and consternation of the game's developers . . . The players, in other words, started to behave like citizens anywhere" (Kim, "Killers Have More Fun," 143.)

20. Murray describes a player complaining about the Clone Wars arcade game, "why should I want to kill these guys? . . . We should all be working together" (*Hamlet on the Holodeck*, 52). The story of the game involves fighting through guards to attain an alliance against a larger enemy. Murray also discusses the effect of switching sides in Mortal Kombat or fighting for the evil Empire in one of the Star Wars games (147).

21. Espen Aarseth, *Cybertext: Perspectives on Ergodic Literature* (Baltimore, Md.: Johns Hopkins University Press, 1997), 145.

22. It is worth recalling that in its original application to media the image of surfing was meant to describe balanced awareness and skilled navigation through dynamically changing conditions. The original use of "surfing" in this context is claimed to be in the essay linked from <http://www.netmom.com/about/surfing_main.htm>.

23. John Cage, Landscape no. 4 (March no. 2) for 12 radios (1951), and Radio Music, for 1–8 radios (1956).

24. Because the modes of critical inhabitation I have been discussing do not reformulate the immersive artifact in terms of another given discourse, but do require attention to the structure and movements within the artifact, they involve new literacy skills applied to new media. However, they are not just investigative techniques for disengaging a message; their sensitivity to processes and goals includes finding ways of intervening within the overreachings and mixtures involved in establishing the artifact's unities and activities.

25. An article by the politically influential science advisor Vanever Bush is generally accepted as the first adumbration of the notion of hypertext linkage. See "As We May Think," *Atlantic Monthly*, July 1945, 101–8.

26. The difficulties with maintaining a distanced critical stance in large hypertexts are discussed by Stuart Moulthrop and Nancy Kaplan, "They Became What They Beheld: The Futility of Resistance in the Space of Electronic Writing," in C. Selfe and S. Hilligoss, eds., *Literacy and Computers* (New York: Modern Language Association, 1994).

Digital Hybridity and the Question of Aesthetic Opposition

Johanna Drucker

University of Virginia

I S there a fanatical inevitability in the capacity of electronic media to absorb all forms of human expression and experience into data formats? In his discussions of aesthetics and ideology, Theodor Adorno continually reiterated the caveat that when positivist logic invades culture to an extreme then representation appears to present a "unitary" truth in a totalizing model of thought which leaves little room for critical action or agency.[1] Pulling this unity apart is essential to critical rationality (as distinct from instrumental rationality) in its struggle to maintain a gap between data and idea, form and experience, the absolute and the lived. In the hybrid condition of the digital the separation necessary to sustain the distinctions between the instrumental and the critical appears to be precluded. The "absolute" identity of the mathematical underpinnings of all digital activity seems to collapse concept and materiality into one and the same as an encoded file. As the popular idea of technological "truth" continues to function as an instrumental force in the increasing rationalization of culture, artwork which renders such "truths" assimilable, performing in what Adorno would term a reconciliatory manner. Such work seems profoundly insidious—unless it can be qualified within a critique of its assumptions, claims, and premises.

Late-twentieth-century technological innovation pushes the boundaries of once discrete areas of cultural activity, transforming an ever increasing number of arenas into colonized domains within the managerial bureaucracy of data processing. As it does so, the hard and fast opposition between two traditions which had been markedly distinct in the visual arts are increasingly difficult to sustain—in part because they each depended upon identification with contradictory concepts of the role of reason. The first is the antilyrical, antisubjective, rational tradition of art which aspires to the condition of science. The second is a humanistic, lyrical, subjective romanticism which opposed emotional, natural, and/or chaotic forces to those of technologically driven progress. Hybridity of machine/ organic entities is a current condition. The old oppositions no longer hold—the machine is the flesh, the body is technological, nature is culture. The cyborg is the sign and actualization of the current lived condition of humanity—and more and more it is clear that as the arts have helped familiarize and legitimate many of these once unthinkable ideas the capacity for aesthetics to regulate boundaries between rational and irrational regimes of technology is limited.

There are many areas of contemporary art activity in which these themes could be examined: the imagery of mutation, sculpture and installation work which merges new technology with conventional media, work which extends the human body through technological prostheses or otherwise toys with machine aesthetics in new, synthetic ways.[2] All are interesting manifestations of a profound transformation. But the implications of this change can be brought into focus through a more narrow, and perhaps more fundamental avenue: an inquiry into the identity of digital technology. Digitalization seems to offer the possible fulfillment of Gottfried Leibniz's dream of a universe fully available to logical analysis and description—in short, his vision of *mathesis* as a complete logic that maps in a one-to-one correlation between thought and representation.

As art's dialogue with technology extends into the digital arena, the questions that arise can be posed in philosophical terms which frame their political implications. Here these questions will be posed in terms of the work of Adorno's conception of aesthetics as potentially resistant (no longer liberatory, given his profound pessimism). At midcentury, Adorno struggled to articulate the capacity of aesthetics to resist the forces and tropes of instrumental reason informing an increasingly commodified consumer society. All cultural production (and reception) had come to mirror the processes of capitalism, with their mind-dulling repetitions and formulas, while the capacity of what passed for reason to perform with destructive force had been made all too vividly clear within the events of the Second World War. Aesthetics could only effectively resist such processes through a refusal of utility and systematicity. Adorno placed considerable weight on formal strategies of artistic production as a means to achieving this goal. These can be identified by the terms *determinate irreconcilability, dissonance,* and *nonidentity* within a work of art.[3] Are these concepts sustainable within the context of the digital production of works of art? Or does the qualified absence of materiality of the digital environment fundamentally alter the way an artwork's identity and self-identity can be conceived within the dialectic of subject/object relations which Adorno derived from Hegel's aesthetics?

A secondary aspect of this discussion is the relation of twentieth-century art to positivistic logic and its embodiment of an Enlightenment ideal of reason. Logic only rarely serves as the thematic subject of a work of art, but positivist assumptions about systematicity underlie the motivation of many twentieth-century artists' practices. Art using digital technology depends on mathematical logic and its peculiar hybrid sibling, programming language. Within digital technology an image or artwork comes to be perceived as "information" and therefore as an essential, absolute, quantifiable identity. The notion that an image can be "reduced to" or is "equivalent to" a data file, an algorithm, a program, or any mathematical, quantifiable identity are all ideas which give rise to a notion of digital identity as absolute and certain.

Leibniz's dream is Adorno's nightmare—and Adorno and Max Horkheimer were quite clear on this point in *Dialectics of Enlightenment,* when they suggested that Kant's aesthetics of purposelessness is a necessary (the only) antidote and means of resistance to the domination by all sections of society and culture by the enslaving, deceptive, forces of the mass culture capitalism (always read as an extension of rationalized modes of production into the cultural sphere). In citing the ways such a technorationality infused itself into cultural practice, Horkheimer and Adorno take issue with the ways the perceptions of representation are themselves subjected to such a positivist logic. This occurs, Horkheimer and Adorno suggest, in the approach to language/image in which a literalist, positivist interpretation forces representation into a collusion/elision with the "real" so that one takes the "word for the thing"—the "image for the real"—the "representation for the referent" in every instance.[4] This collapse of the discrete structures of representation into a perceived unity makes it virtually (in the technical sense of the word) impossible to insert any critical distance into an understanding of representation and its social function. The distance between Leibniz and Adorno coincides with the temporal span which encompasses cultural modernity. What was for Leibniz the glimpsed possibilibity of total, all-encompassing, descriptive and analytic reason has become for Adorno (and Horkheimer and others) the nightmarish image of totalizing control effected by instrumental rationality.

Artistic engagement with modernity as a phenomenon which rapidly transformed every area of cultural production, communication, and administration attached a wide range of valences to rationality and technology. A keen awareness of the effects of the radical changes wrought on lived experience produced celebratory as well as critical responses. The positive utopianism of avant-gardes was premised on the belief that progress promised liberation of humanity from oppressive labor and its social constraints, while the nihilistic negation of technorationality attacked these premises and the entire tradition of positivist thought with its origins in the Enlightenment. There was also a tension between the idea that the escape from reason was the means to salvation for the human spirit and a contradictory position that invoked rationality as an antidote to injustice and social inequities in the name of an Enlightenment project. Many reforminst and counterculture movements of the late nineteenth century give evidence of increased momentum for a critical opposition to an unchecked and unquestioned concept of "progress" as the automatic gloss on any feature of modern technology or its effects.

By the early twentieth century, aesthetic engagement with the technological manifestations of rationality (and later the irrational manifestions of supposed reason) map across a considerable range of more sharply defined positions. These include the enthusiastic production of a machine aesthetic—Robert Delaunay's renowned paintings of the Eiffel Tower, Fernand Léger's machine motifs, and futurist works (and rhetoric) in praise of motor cars, trains, and industrially produced objects. A clear privilege was assigned to the inorganic and technological over and against the holistic, humanistic, or organic. The ironic counterpart in the work of Dada poets and painters, conspicuously nuanced in the oeuvre of Marcel Duchamp, is an understandable contemporary counterpoint response. The embrace of chance operations and resolutely asystematic negations of the very premises of rationality eschewed any nostalgic return to the lyrical, personal, or artisanal which had been held out as antidotes to industrialization and modernity in earlier decades in the

Arts and Crafts movement, in symbolist aesthetics, or among the Pre-Raphaelites.

The influence of reason in the spheres of cultural practice can also be seen in the colonization of humanistic "soft" disciplines by a "scientificizing" attitude. The idea of the social *sciences* of linguistics, anthropology, sociology, and politics gains currency in the late nineteenth and early twentieth centuries. The aspirations of the visual arts to identify their own systematic tenets and become practiced with guaranteeable, repeatable, and predictable "results" are part of this general cultural pattern. All of this would be merely a historical description of a past cultural moment were it not for the fact that the continuum which stretches from this scientificizing impulse continues into the present, providing assumptions on which the legitimacy of the digital is premised. And one crucial factor in forging the connections which interweave the sciences and humanities is the historical emergence of logic (first in George Boole's "algebra") which comes to serve as the basis of programming languages. Logic comes to dominate the philosophical inquiry into language in the work of anglo-american analytic philosophers Bertrand Russell and Alfred North Whitehead, and the group around Rudolf Carnap. On the basis of this link the concept of *mathesis*—the logical calculus of thought—revives its potency within the digital arts—since all digital and electronic operations are based on a mathematical processing (if not an algorithmic production) of visual images and works of art. With the realization that it is such mathematical logic, and not a machine aesthetic, which is the basis of digital work's relation to the technological as a concept and as a practice, it is possible to begin to place this link in the context of the larger discussion of those overt connections between art and machines, aesthetics and technology, in the twentieth century. The distinction between a machine aesthetic and a more general aestheticizing of technological processes within the celebration of logic and systematicity can be elicited from the different premises exemplified by Lazar El Lissitzky, Fernand Léger, and Wassily Kandinsky.

In a 1923 work titled *The New Man*, El Lissitzky depicts a single figure, sufficiently anthropomorphic to clearly fulfill the expectations of its title, but composed of forms constructed entirely with mechanical drawing instruments. A single circular arc describes the curve of the torso from shoulders through hips. The limbs are stretched outward in the directions of the four axes of an axonometric diagram, their volume thinly designated by minimally described shapes. A hand is described by a thin arc segment of a dark circle, another is merely the terminus of several aligned rectangular strips, the feet created out of short boards resembling the base of a lamp-stand crudely constructed from carpentry remains for purposes of utility. The head, rising above the squared-off geometry of the shoulders, is made of two partial circles, overlapped along a central gap between them in which features of a cut-out profile are sharply suggested. There is no simulacral imitation of a machine in this work. Unlike the ironic mechanomorphs produced by Francis Picabia in the same period, with their satiric play on the erotics of machine aesethetics, Lissiztsky's image makes a straightforward claim, placing the constructed aesthetic forward as an unequivocally positive vision.

But Lisstizky's image only makes use of mechanical instruments in its construction—it is not itself an image of a machine. The integration of elements in the figure do not articulate as working parts of a geared, motored, or functional apparatus. The image is not of a man-as-machine. Nor is it a simplistic mapping of the features of a machine form onto the figure of a man. The work takes its point of aesthetic departure from a more fundamental integration of organic and machinc principles as the basis of design. The merging of the two takes the form of a hybrid aesthetic in which the assumption of a shared system underlies both modes of production. *The New Man* is not a machine, but a systematic production. There is no identifying detail, no sense of a portrait, no sense of an identity within the image—it is generic, prototypical, and emblematic. It reads as a sign and program, a condensed image signifying a prescriptive rather than descriptive image of a "new man." Like a template or pattern, the geometric forms suggest the possibility of manufacture and production, not depiction, in their spare simplicity.

By contrast the work of Ferdnand Léger, *Ballet mécanique*, places exaggerated emphasis on reconceiving the human body in terms of a single metaphor of machine imagery. Léger, whose early cubist work of the decade before World War I was largely engaged with the analysis of forms in a reductive geometry, had become infatuated with machines during his war experiences. "I was dazzled by the breech of a seventy-five millimeter gun which was standing uncovered in the sunlight: the magic of light on white metal."[5] Apocryphal though this statement may be, it exemplifies the all too familiar and now much discussed relation between the rationalist aesthetic of certain early avant-garde movements and the militaristic emergence of fascism with its tropes of efficiency. The negative characterization of instrumental rationality articulated by Adorno attaches with an understandable vehemence to the aestheticization process which enables such values.

The superficiality of Léger's grasp of machinery is evident in the rapid reproduction of hard-edged, shiny, machined forms as the basic building blocks (and surfaces and treatments) of his work. Even the phrase "machine aesthetic" seems from the outset to be stigmatized by a superficiality and a sense that it is a mere act of appropriation which combines the two in a casual fashion parade of formal novelties decked out for consumption. The "machine aesthetic" can be read as a link between the industrial machinery of overproduction and the need to provoke consumption through tempting the appetite of the consumer. Such aestheticization is also a response to the early twentieth-century ennui, a neurasthenic and general cultural malaise in negative response to increased industrialization iconically signaled in Charlie Chaplin's *Modern Times*.[6] Such malaise may not have resulted in a full-blown critically articulated opposition, though sporadic and organized resistances occur throughout the industrialization of nations practically from the moment of the inception of industrial methods of manufacture.

Lissitzky's work shares an embrace of the rational with that of Léger, but it exemplifies a set of ordering principles rather than merely celebrating industrial production and machinery in its many manifestations. Inherent in his concept of systematicity is a vision of individual integration into a social whole as a positive possibility and the basis of functional collectivity. Thus the conceptual parameters of Lissitzky's approach to systematic thought are not burdened by the negative critical characterization which Adorno will attach to them in the later 1920s and then from that point onward. But the systematic character of Lissitzky's aesthetic is still at a remove from the overarching concept governing the longer historical argument—namely the argument about the relation between *mathesis* and aesthetics. Lissitzky's distinctive approach to design is clearly consistent with a more general agenda of constructivism and, in its grounded relation to instrumental application within a social realm of real production—the basis on which such an aesthetic developed, whatever its roots within more esoterically inclined abstractions—has to be factored into any interpretation of its forms.[7] This is useful since this historical evidence justifies the critical link between aesthetic and social/cultural implications of systematicity, but it also distances Lissitzky somewhat from Kandinsky, whose work is a useful third term to introduce into the discussion of rationalized aesthetics in the early twentieth century.

Although Kandinsky loads his earlier searches for a systematic "language" of visual symbols to a spiritual agenda redolent with synaesthetic dimensions, his work in the 1920s establishes certain paradigmatic texts and approaches. *Point and Line to Plane,* published in 1926, was a crucial articulation of an abstract vocabulary of forms and their relations. The project of this work is to establish terms of certainty, of predictable and guaranteeable formal relations, among visual elements. In short, the premise that such a possibility can be envisioned is itself a product of the more general early twentieth-century embrace of a rationalism, and the execution is possible because the general understanding of how such a project could be carried out has matured. Kandinsky's work, naive as it may appear in retrospect, established a precedent which had a developed following in the realm of graphic design. Few artists' handbooks of the *Point and Line* variety have the credibility of Kandinsky's early endeavor, though the writings of Gyorgi Kepes, Johannes Itten, Josef Albers, and others are all continuations of this rational approach to visual experience and representation. Many lesser titles in this vein deteriorated into copy-books of simplistic formal solutions or how-to pamphlets without critical or theoretical aspirations. But in the realm of graphic design the lineage is considerable and the idea of graphic systems became a useful instrument in the increasing corporatization of design functions as well as in the later field of information design. The *Catalogue Design Progress* (1950), Ladislav Sutnar's elaborate catalogue of templates for designing legible hierarchies and sequences of display, is an outstanding example of a genre which includes the full range of work from trade handbooks to sophisticated and elegant models.[8]

Kandinsky's work provides another key element—the suggestion of a Pythagorean faith in alignment of values of color into form into mathematical number. Kandinsky's faith in the absolute, in "value" as an essence feature of form, combines the mystical and mathematical without apparent conflict. The idea of synaesthesia is closer to the idea of the mutability of the digital file than any idea of aesethetics wedded to materially apparent form would be—since there is an implication that the numerical value of a color, a form, or a musical note resides in an essence outside its material. In digital storage, a file has no necessary relation between the form in which it was input (as a text, a sound, an image) and the form of its output. The contrast among the three artists' work allows distinctions to be made clear between image and representation at a conceptual level. The Léger, with its machine motifs, is finally an *image of,* or *a representation of,* superficial and stylistic, whose referent is outside itself no matter how much its own surfaces and their treatments replicate the modes and manners of industrial production. It poses no critique since its aestheticization serves to reconcile such production with the culture as "natural"—as the leitmotif of the times. Lissitzky's *The New Man* enacts its aesthetic premises within the image, its premises, and its production. Still celebratory, its own integration is complete. The subjugation of natural material of human life and form into the modes of rational production has hybridized the two domains indissolubly. Machinic and systematic conception are one with the anthropomorphic; criticality has no space in which to enter; the aesthetic naturalizes the activity as an effect. Kandinsky's systematic, mystical equivalences abstract value from form while asserting the absolute capacity of form to embody this value in an immutable and uninflected presencing of truth. Value is form as truth in essence in this equation with its many interchangeable terms. Within the consideration of *mathesis* Léger's work falls away as mere skin, a surface effect, while Lissitzky's work exemplifies a set of structuring principles, and Kandinsky's a revelatory mystical faith in number as essence. Kandinsky never suggests a mathematical calculus of form or composition—he merely implies it at every turn within the *Improvisations* he painted and their musical/mathematical analogies. All three embrace, consciously or not, a positivistic and scientificized attitude toward art production and each participates in the early twentieth-century engagement with systematicity in the visual arts. But in the early twentieth century it is clearly not machine aesthetics which is the most complete expression of rationalism; rather it is the fundamental attempt at grounding art in a basis which approaches science, mimicking its capacity for logical certainties, which are the goal of this positivist imperative.

Midcentury developments in the dialogue of positivism and art are as varied as those of the earlier decades and are tempered considerably by the post-World War II reflections on the destructive character of technorationality. Information science, which flourished within the adminstrative and military industries of the war years, expanded out of the restricted domains of esoteric application extending the ever broadening reach which would, ultimately, push it toward its late twentieth-century penetration of every aspect of contemporary culture. The rational surfaces of modernist grids, replayed in the minimalist canvases and structures of system-generated art, give ample evidence of the persistence and aesthetic potency of this aesthetic. Sol Lewitt's famous dictum, "an idea is a machine that makes art," sutures the machinic and rational within his aesthetic—no matter how qualified each of these terms must be in actually assessing his projects.

Computers began to find their way into art projects in the 1950s and 1960s, usually in elaborately conceived works whose conceptual parameters required a considerable amount of input for a relatively piddling material output. The burden of production was such that this work has a high coefficient of conceptualization in proportion to the quality of final product. But such work established the basic paradigm distinguishing input from output, idea (algorithm, program) and material (printout, template, form) in a highly mediated manner—so that the "stuff" of a piece is data to be manipulated through process. Fundamental to conceptual art as well as computer-generated work, this paradigm signals a radical break within the aesthetic underpinnning of the fine arts since it renders overt the terms on which idea and material are distinguished—or cease to be distinguishable, depending on the extent to which the very identity of a work of digital art is posited within its digital file.

There are many works within conceptual art narrowly defined (and within the reconceptualization of art broadly understood) which take the paradigmatic premises of this distinction between idea-as-program and form-as-mere-incident-of-output as their basis from the 1960s onward. There is a corollary to be drawn between the classical sense of "essence" and "accident" here—in which the digital file has an immutable identity and the individual outputs it can be used to generate are each mere and unique but ever so slightly debased and individuated instances of that original. The question is whether the "essence" of the digital work is in fact its file, encoded and encrypted and clearly mathematical, or whether the digital work resides in a fulfilled expression of that file into a material form depends on the conceptual parameters of the piece.

Here it seems useful to come back to a framework provided by Adorno in the following phrase: "The Same, which the artworks mean as their what, becomes through how they mean it, an Other."[9] Adorno's aesthetic agenda depends on a separation between subject and object, on a persistent and irremediable difference at all levels of artistic production, which allows criticality to function. This "difference" takes the form of keeping idea separate from material—the "same" which is idea or content or thematic problem is made into an "other" through the extrusion into and embodiment in a materiality. The inability of these two to be identical can be maximized in works which promote a dissonant, nonunified, nonunifiable condition in their formal realization. Such work, by its own form and by its disruptive place in the cultural landscape of otherwise too-easily consumable objects, serves to keep the critical function alive. Dissonance without closure, difficulty without reconciliation ("determinate irreconcilability")—these functions keep the rational from dominating the material, keep reason from turning nature into its perfect image as some mere imitative form.

Two examples of digital production provide contrasting problems in extending Adorno's premise into an analysis of the digital work: *Evolution of Form* (1990) of William Latham and the virus works of Joseph Nechvatel. Latham's work consists of algorithmically generated forms. They grow, mutate, and adapt within the parameters of an also mutating program. Its operation translates the problems of evolution into a reworked solution at the level of the program—the forms evolve according to the various solutions. The images are organic in appearance—tiny units of pale fleshy tissue emerging like some kind of mutant growth. They look more like brains, ovaries, or internal organs than like self-sufficient organisms—clusters of cells organizing in order to perform a specialized rather than autonomous function. As forms, their identity is entirely linked to—is even, on some conceptual level isomorophic to—the files generated through the mathematical operation of the algorithms evolving and acting on each other. As an image of the process of form-giving, the work raises questions about whether the file is one with the image or distinct from it. The image cannot exist without the file—except in secondary format as photograph, printout, or other hard copy manifestation. But that level of existence is as secondary as a photographic reproduction of the *Mona Lisa*. As an original, the *Mona Lisa* is not linked to a file as its essence: it is its own essence and has its own ontological integrity—it *is* in every sense in its material manifestation. But are the "how" and the "what" of Latham's work distinct entities—or even distinct components within the original image-generating-file and file-linked-image? Is it even correct to name these as two different elements? Or are they merely two aspects of a single entity? And whether they are a single entity or not, they share a common premise—that the fundamental ontological condition of the image is as a mathematical entity. This entity is utterly consistent with the notion of a logical proposition, one which is able to be absorbed into the language of certainty and guaranteeability essential to logic. There is a lurking sense that the absolute truth resides in this file, in the digital code.

Nechvatel's virus-affected image files offer an interesting contrast. Nechvatel creates image files—graphic, pixel-based images created through scanning, through the upload of visual data into a pixel tapestry. Such images are not algorithmic in their form or creation— they are stored as encoded descriptions of a surface or area of display. There is integrity to the relation between image and file, but only descriptively, not prescriptively. The graphic quality of such images means that they carry more data than can be described by mathematical formulas for their forms and shapes—merely carrying it as incidental information which has been inscribed into code storage (in the same way that a photographic surface carries much more information than can be described in a linguistic or algorithmic prescription for its forms). This stored file is then subjected to a virus which Nechvatel introduces electronically, getting it to "work" on the image file. Nothing happens on the display screen while this occurs. The virus works its way through the data files, damaging them, reconfiguring the pixel tapestery. When the process is complete, the new tapestry is displayed. But is the relation of the new display to its data file the same as the original? Yes and no—yes because both are "mapped" as display, merely in a descriptive sense, but no because the second image/display also contains the visual signs of the algorithmic, programmed, viral damage. The image is not merely one with the generative file: it is one with an incidental stored file and a secondary process which has resulted in an altered file. The ontological condition of the altered file isn't the same as in the case of Latham—where the image and file evolve simultaneously. In Nechvatel's case the display is after the process, and is a record of it, and suggests a potential recuperation of that transformative process in the altered form of the display. The Nechvatel image does not have the same unity as the Latham—it may have the same relation of identity between final, extant file and display, but it is not unified in the originary sense which the Latham file is. All digital files have this unity of file and display, but within the record of their production processes there are varying potentialities for distinguishing critical distinctions. The idea of Nechvatel's virus works on the material of the digital file—and in that gap the critical enters and produces a dissonant fracturing. Nechvatel's image contains a critical opposition to itself, as part of itself, a split identity in its apparent unity. This is because Nechvatel's graphic file preexists the mathematical operation—thus allowing the "stuff" of a material image to enter into a relation with the immaterial action of a mathematical operation. Nechvatel's is not a hermetic universe, singular and closed, in its relations of ontological essence and manifestation.

The virtual is not the absolute—not some fused condition of unity in which there is no conceivable critical distance. But the virtual can enact a new condition of unity which precludes, by simulacral effect, by a confusion of image and the impossibility of difference within it, any substantive difference between concept and matter, idea and form. If the absolute is the condition of giving up of the distinction/difference between subject and object, image and thing, representation and referent—then the virtual is rather a forced admission that there never was a difference except in theory. The virtual does not represent, and also *is* not, an achievement of unity after a struggle. This seems to be the case with Latham's *Form*.

The positivist underpinning of artistic scientificization gains considerable justification in the digital—where the "materiality" of art production is inscribed and contained entirely within the mathematical equivalences of electronically stored files. The question is whether the digital image is completely self-identical, hopelessly constrained within the terms of a unity of idea and form at the level of trace inscription in binary code, or whether the gap which separates file from each and every instance of manifestation in material output (display on the screen, print, pattern, template) makes the separation of concept and execution greater than ever before. One can take the position that the file is the work in the most absolute ontological sense of its being the essence of the work at a level of descriptive code which replicates without variation outside of temporal/historical/subjective influences. Or one can take the opposite position, that the code is utterly ephemeral, only an effect of highly mutable, subject-to-fashion-whim and direction of the ongoing industry changes, and that the relation between "idea" as code and "form" as object/material manifestation has never been more clearly marked than it is in the discrete boundaries of file and form, encoded information and material manifestation.

Does *mathesis* have its triumph within the realm of the digital? Within digital is there a gap preserved or a gap closed as if it had never been open between "natural" and "computational," the serendipitous and the systematic, the organic/replete and the synthetic/constructed? If *mathesis* describes the ideality of form—that is, form which is apparent to consciousness without any need to be manifest in material—then the digital condition of code storage fits that description as well. But digital imagery is also the curious hybrid of materially inscribed form (massively dependent on the apparatus of production, the computer, terminal, magnetic storage devices, discs, and drives) and immaterially apparent form whose two aspects are inseparable from each other. The coming-into-form which involves a splitting off from idea into matter is circumscribed within a single operation in the digital environment in a work like Latham's—but not so clearly in the case of Nechvatel. Each poses a different problem with respect to the self-identity of digital images and the dialectics essential to critical intervention. To frame it another way, in a paraphrase of the Adorno quote above, in certain instances the *how* of the digital environment can be precisely the same as the *what*. The means and matter of expression of form/content, idea/expression, essence/accident can become indissolubly unified in the mathetic condition of digital storage, thus questioning the possible basis on which aesthetic opposition can even be premised, let alone sustained.

Notes

1. Hauke Brunkhorst, "Irreconcilable Modernity: Adorno's Aesthetic Experimentalism and the Transgression Theorem," in Max Pensky, ed., *The Actuality of Adorno* (Albany: State University of New York Press, 1997).

2. The work of Mark Pauline and the Survival Research Lab, performance and robotic artist Stelarc, video artist Alan Rath, sculptor Janet Zweig, and the collaborative work of Heather Schatz and Eric Chan, to name a handful.

3. See Peter Osborne, "Adorno and the Metaphysics of Modernism: The Problem of a 'Postmodern' Art" and Peter Dews, "Adorno, Poststructuralism, and the Critique of Identity," both in Andrew Benjamin, ed., *The Problems of Modernity: Adorno and Benjamin* (New York: Routledge, 1989).

4. Brunkhorst, "Irreconcilable Modernity," 52.

5. Max Kozloff, *Cubism/Futurism* (New York: Harper Icon, 1973), 40.

6. Stephen Kern, *The Culture of Time and Space* (Cambridge: Harvard Univeristy Press, 1983).

7. Stephan Bann, ed., *The Tradition of Constructivism* (New York: Da Capo, 1974).

8. Richard Hollis, *Graphic Design: A Concise History* (London: Thames & Hudson, 1994), 118.

9. Brunkhorst, "Irreconcilable Modernity," 52.

The Cyborg and the Net: Figures of the Technological Subject

Silvio Gaggi

University of South Florida

A T this moment in the evolution or devolution of the postmodern, technological subject, two figures have emerged that
evoke the dreams and anxieties associated with an imagined future
into whose reality we are rapidly rushing. The *cyborg* and the *net*: the
cyborg, an incarnated, material, entity, a bricolage constructed of
organic and inorganic, natural and artificial elements; the network,
a conceptual, electronic space navigated in a way analogous to the
way one finds one's way through an unknown city, struggling to
build a conceptual map, getting lost and confused, blissfully or
frightfully decentered, like playing in a labyrinth whose organization we fear we will or will not master. These two figures recapitulate
an ancient binarism: the material, the carnal, the real; the mental,
the spiritual, the ideal. But the distinction between them is not always so clear—a cybernetic prosthesis is a virtual limb; the Internet
is a mental prosthesis—and their implications spill one into the
other in many ways, as well. These figures, which recur in the artistic
production of the contemporary world, are also realities that we experience in the technology of everyday life. They embody both the
positive and negative hopes and possibilities we experience as we
enter the new millennium.

In a sense, humans always have been cyborgs because they have
always used tools to supplement and extend themselves, physically
and mentally as well. However, at this point technological supplements have become so ubiquitous, the relationship between the
human and her tools so intimate, and the line between them at
times so seamless, that a new entity, one that transgresses categories
of human and technological, has emerged. It is not just a matter of
using tools to extend our "selves." It is a matter of becoming "one"

125

with our tools. But this is a "one" that is fluid and provisional, a "one" constructed of parts that can be deconstructed and reconstructed, that can morph slowly or quickly into different "ones." It is a one that, because it is so obviously provisional, challenges the notion of a permanent and unitary subject and provokes us to doubt the existence of that subject. Or, at least, to attempt carefully to locate whatever residue of an abiding subject there is that may remain, to look for the "ghost in the shell" (to appropriate the title of the Japanese *animé* film in which cyborgs figure significantly).

For Donna Haraway the cyborg is a central reality and metaphor for the subject in the contemporary world. The body is extended and the skin is penetrated by inorganic, technological supplements that create a hybrid subject that transgresses conventional categories: categories of gender and race as well as the categories distinguishing machine and human. In her "Manifesto for Cyborgs" she writes, for example, that " 'women of color' might be understood as a cyborg identity, a potent subjectivity synthesized from fusions of outsider identities," and she identifies three broad categories of boundary transgression that are signaled by the cyborg: the boundary between the human and the animal, the boundary between the organic and the machine, and the boundary between the physical and the nonphysical.[1]

The technological penetration of the body may be experienced as a violation, a kind of technological rape; the history of technology as a tool for violation, incarceration, surveillance, and psychological and social invasion hardly needs to be argued. But technology need not be used this way, or at least that is the hope. Many writers and activists are struggling to identify strategies for turning technology into a tool for checking hegemonic power and enlarging rather than limiting freedom. A collection of essays entitled *Technoculture* describes various such strategies, utilized by individuals involved in activities such as computer hacking, rap music, and alternative video.[2] Haraway describes a strategy of working in the "belly of the monster,"[3] of relinquishing the dream of change brought about by agents working from an "objective" position outside the system and instead entering the system of late capitalism in its high-tech mode and altering that system as an agent from within, surrounded by it. Embracing technological supplements and constructing oneself as a hybrid cybernetic entity might be a choice of intelligence and will, an incorporation of the technological into the organic self, creating a new self—or, for Haraway, the possibility for all sorts of new selves.

Haraway is interested in the consequences and possibilities this cyborg world has for women and feminism, in particular. Politically the cyborg seeks provisional alliances based on ad hoc groupings of mutual interest, alliances that may, in different situations, be organized around or crossover boundaries of race, gender, or class, none of which constitute essential categories or essential bases for permanent union. Nor would the distinction between humans and animals, any more than that between humans and machines, be regarded as an essential one. A "cyborg world might be about lived social and bodily realities in which people are not afraid of their joint kinship with animals and machines, not afraid of permanently partial identities and contradictory standpoints." "The cyborg is a kind of dissassembled and reassembled, post-modern collective and personal self."[4]

Haraway argues that the cyborg, being a construction, is parentless:

Can you come up with an unconscious that escapes the familial narratives; or that exceeds the familial narratives; or that poses the familial narratives as local stories, while recognizing that there are other histories to be told about the structuring of the unconscious, both on the personal and collective levels.[5]

Thus there is no nostalgia for an earlier era, an ancestral past or original innocence or wholeness for which one can yearn. This may bring about a degree of existential alienation, to be sure—after all, one's knowledge of one's "roots" can provide a sense of identity and connectedness, but the absence of such roots also provides freedom from the need to seek an original, true, "self" to which one may subsequently be bound. Being parentless and, consequently, lacking a definition of the self that is given at the outset, a priori, the individual has no patriarchal symbolic around which to construct an Oedipal subject and no maternal semiotic to abject (to adapt the family myths of Sigmund Freud and Julia Kristeva). The constructed nature of the cyborg is foregrounded at the outset, and deconstructing it, clarifying the artificial nature of what might have appeared to be natural, is hardly necessary.

Cyborgs have become stock figures in contemporary science fiction and cyberpunk fiction and film. Molly, in William Gibson's 1984 novel, *Neuromancer*, which defined the cyberpunk genre in literature, is a cyborg, replete with implants, sensory supplements, and technological extensions, and she is a major prototype for all female cyborgs to follow. Not only is she a human-technological hybrid, her identity is also a collage of diverse other aspects—she is a "hit man," a "ninja," she was at one time in her life a prostitute (or "puppet," to use the novel's own cyberpunk jargon), and, having a strong aggressive impulse and being physically superior to Case, the novel's protagonist, she is a femme fatale.

From the standpoint of gender representation, Molly, as an image of Haraway's technofeminist emancipation, is only skin (or implant) deep, however. Underneath it all, Molly is a woman, with all the vulnerabilities that traditionally come with that traditional category. From this perspective what is most significant is that she is really a technological variant on the classic femme fatale, a sexy, dangerous, female. She happens to be a cyborg, but that only makes her more exotic and exciting: if she doesn't love you, she just might kill you; or maybe she'll do both, love you and kill you. The danger she poses enhances the excitement of the conquest.

Moreover, in terms of the nature-culture dichotomy, Molly, in spite of her technological supplements, does come down on the side of nature, with Case on the side of culture. Molly occupies the material world, the world of the body, the world of "meat," to use the novel's term for the material world, as opposed to the immateriality of cyberspace. Molly explains her aggressive tendencies by saying, "I guess it's just the way I'm wired."[6] Here the artificiality of wiring is conflated with an essentialist notion of "nature." Molly's phrase might just as well be, "That's just the way I'm built," "That's my nature," or even, "It's in my genes." Whether she's programmed electronically or genetically makes little difference in determining what she is, which she accepts as if it is something inalterable that lies outside of her will.

Case, on the other hand, is a "console cowboy," a virtual reality hacker who disdains "meat" (whether the meat is biological or technological), regards it as an unfortunate necessity at best, and is most at home in the artificial "reality" of cyberspace, a realm that is, if not platonic and intellectual, at least nonmaterial and separate as much as possible from the limitations of the flesh. Thus, even though the novel on certain levels of its action, in much of its inventive imagery, and in many of its characters, dissolves the opposition between nature and culture, human and artificial, biological and technological, on a very important level, that of the relationship between Case and Molly, the traditional dichotomy between male and female, with the male assigned culture and the female assigned nature, is maintained.

Indeed, much of the novel suggests that, in its technological, dystopian future—which though dystopian does have a colorful, attractive, exciting aspect—there is a rejection of the body and what Kristeva would call the "abject" that is associated with the body, especially the maternal body.[7] As with Haraway's cyborg, the characters in *Neuromancer* (all of them, not only Molly, in some respect or another, cyborgs) are parentless. They may have histories, but their histories do not extend further back than young adulthood. Case rejects the "meat," though of course he is entirely dependent on it, if only as an instrument that allows him to live and to enter cyberspace. Cyberspace itself is fleshless, meatless. It has its dangers, even life-threatening dangers, but it lacks the tactile, smelly, messiness of nonvirtual reality. And even Molly—physical though she is—is a polished, machined, efficient physicality, a surface of metal, leather, and mirrorshades. Where the flesh shows through she is smooth and perfect. If she were painted it would be done with an airbrush and a pen, or, better yet, it would be done on a computer. Nothing in the novel messes up its smooth, high-tech surface. Even when Molly is injured, she feels pain, but the damage is located safely below the leather and metal that is her supplemental skin.

Other postmodern cyborgs include Kathy Acker's Abhor, from *Empire of the Senseless*, who is "part robot, and part black."[8] The presentation of these two terms as an opposition confuses categories even more than might otherwise be the case: the contrary of "robot" is "human" and the contrary of "black" is "white." "Part robot . . . part black" creates a categorical muddledom not only because lines between categories are crossed but because the lines themselves are confused. The discourse of the novel, similarly, is a polyglot of dialects, mixing language that is "high" and "low," educated and uneducated, slang and formal. Abhor's discourse shifts irrationally from vernacular street talk to the sophisticated jargon of political or psychoanalytic theory. There is also Major Motoko Kusanagi, from the *animé* film *Ghost in the Shell* (*Kokaku kidotai*, 1995), a female cyborg for whom, no doubt, Molly is an antecedent. Like Molly, she shares qualities of toughness and vulnerability, but her vulnerability is more up front, less concealed. A cyborg's "ghost" is the remnant of the organic and human that remains beneath its technological supplements and alterations. Kusanagi, in spite of her supplements and alterations, is indistinguishable from a real woman, an attractive one at that, and she manages to shed her clothes at every possible opportunity, for little apparent reason other than to please the men who comprise a large part of the film's audience. Significantly, the cyborgs of *Ghost in the Shell* are not only physical organic/technological hybrids, but they also (in contrast to Molly in *Neuromancer*) connect with the network, with cyberspace; thus the two major figures discussed in this essay—the cyborg and the net—converge in the cyborgs of *Ghost in the Shell*.

A figure that is closely associated with the cyborg is the replicant. A replicant, unlike a cyborg, is an entity that is entirely constructed; it is not a human that has been technologically supplemented or a synthesis of human and machine. In this sense a replicant is like the robots that appeared in fiction and film earlier in the twentieth century. However, unlike most robots, the replicant, though artificial, is indistinguishable—or nearly indistinguishable—from a human being. In this way the replicant, like the cyborg, challenges the line between human and machine. Usually replicants' similarity to humans is only skin deep, however. They look like people, and they may emulate human behavior, but something is lacking in their affective lives. Often that something is empathy, which is usually presented as a uniquely human quality. More recent replicants, however, are capable of emotion—or they show signs of having humanlike emotions. In some instances they may themselves be unaware of their status as replicants. Memory implants may create false histories for them, which they fully believe and experience in a way that is indistinguishable from the way humans experience "real" memories. Thus, the line articulating the difference between human and machine is further challenged. Replicants become subversive to their human masters—but more sympathetic to viewers and readers—when they transgress their roles as slaves or workers and seek freedom or survival, independent of human control. In their relationships with humans they recapitulate humans' own relationship with God or nature.

The "robot Maria," from Fritz Lang's German expressionist film *Metropolis* (1927), is an early example of a replicant. She is a robot, technically speaking, entirely constructed by the evil Dr. Rotwang, a Faustian scientist gone awry and a sell-out to the corporate controllers of Metropolis. The robot, who is made to appear identical to the saintly "real" Maria, is used as an instrument for stirring the masses to rebellion and thereby providing justification for a repressive reaction by the corporate overlords of Metropolis. The robot Maria does have a subtle "tell" (to appropriate the language of poker players), a slight imperfection in her behavior that gives her away (to the audience, if not to any of the characters in the film)—an eye that blinks oddly, a kind of technological tick, as if the eye has not been properly coordinated with the rest of her construction.

The tells of later replicants are not so obvious. The replicants in the film *Blade Runner* (1982), as well as those in Phillip K. Dick's novel *Do Androids Dream of Electric Sheep?* from which *Blade Runner* is adapted are physically indistinguishable from human beings, and they are nearly indistinguishable psychologically as well. Like other replicants, they are created for the use of humans, and, as with other replicants, their human masters lose control of them. Unlike robot Maria, human control is not lost because the replicants pursue their masters' purposes with too much enthusiasm and success (Maria is so successful in fomenting rebellion that Metropolis is nearly destroyed), but because they develop their own purposes, independent of those their human creators intended. The replicants of *Blade Runner* develop feelings of passion, affection, and perhaps even love. They are also capable of seeking revenge, and they certainly have the desire to survive—in clear conflict with their four-year life span, a built-in fail-safe mechanism intended to limit their abilities to develop human emotions and to rebel. But in spite of this limit, these replicants even develop the capacity to feel empathy and to be merciful, and they are even capable of philosophy—all very "human" characteristics. True, they are brutal in their treatment of human beings, but empathy is still fairly new for them, and they are understandably upset with their lot in life. In any event, they are no more brutal with humans than humans are with them, both in the early deaths humans have built into them and the death inflicted on them when they are "retired" by bounty hunters called "blade runners."

Blade Runner's replicants do not reveal themselves through an obvious tell—the visible twitching of an eye that reveals a mechanical imperfection—but can only be detected through the administration of a sophisticated psychological and physiological examination—a set of specific questions along with observations of changes in the skin and eye—which allows the administrator of the test to identify replicants and distinguish them from humans. This problem is made even more pointed because at least one replicant, Rachael, does not herself know that she is a replicant; memory implants have created for her a borrowed history that she believes to be her own—though she does eventually come to realize the truth. From the replicant's perspective, it is human, or, at least, a kind of entity that shares with human beings all the important "human" characteristics. The protagonist, Deckard, himself a blade runner, is accustomed to retiring replicants callously, as if the act is not one that is subject to ethical considerations but is morally equivalent to discarding a machine when it stops functioning properly. But, because of the relationship that develops between him and Rachael, Deckard himself comes to wonder about the difference between replicants and humans.

The network is another central figure for those who theorize and speculate about identity in a world that is radically mediated electronically, where the textuality one engages is not the text of literature or the text of painting—or even of film or of video, as they have operated so far—but electronic hypertext. The subject who enters that largest network, the Internet, extends itself outward toward an indeterminate electronic and conceptual horizon, extends if not its corporal skin at least its ego skin and joins a collaborative cybernetic intelligence.

Whether or not one might regard this collaborative intelligence as in some sense or another an *entity* (analogous to colonial protozoans evolving into multicelled organisms) is a matter about which philosophers, theologians, and writers of science fiction can speculate. One might argue, nonetheless, that individuals are shaped by the media they engage, and when the conventions of an existing medium undergo changes, or when the major medium they engage itself changes, individuals' sense of self, as well as of the shape and limits of the self, changes as well.

"Classic realism" in both visual and verbal art has the effect of confirming the viewing or reading subject as a separate and independent locus of consciousness, clearly positioned in time and space. Renaissance painting, through the laws of optical perspective, implies a viewing subject that is localized at a particular point at a particular moment. Jacques Lacan argues that a visual image of bodily coherence is mirrored by a subject that views that coherence.[9] The coherence offered by Renaissance painting, which "mirrors" the visual world, is similarly reflected in the optically positioned subject viewing that coherence. Verbally, a subject may be identified and called to account for itself—"hailed" and "interpellated," to use Althusser's terminology.[10] Classic realist fiction verbally positions the reader by constructing a clear hierarchy of voices with a highest-

level voice that "speaks" to the reader, "interpellating" the reader
and inviting the reader to enter a social and symbolic order that the
reader shares with the narrator.[11] And, of course, Lacan argues that
one only fully enters society and subjecthood when one enters the
symbolic realm of language.[12] Modern literature and art, however,
which often create complex and contradictory points of view and
ambiguous perspectives, have the effect of destabilizing and dispers-
ing that subject rather than confirming it. Different narrating and
reading identities are implied, and viewing subjects are spread
across space rather than localized.

Moreover, when the dominant medium individuals engage shifts,
changes in the psyche are produced. Walter Ong writes: "Since at
least the time of Hegel, awareness has been growing that human
consciousness evolves . . . Modern studies in the shift from orality to
literacy and the sequels of literacy, print and electronic processing
of verbalization, make more and more apparent some of the ways
this evolution has depended on writing.[13] According to Ong, literary
culture is characterized by a more complex, sequential, and hierarchical style of thought than the highly patterned and repetitive
thinking of oral cultures. Enumerations and aggregations characterize oral thinking, but more complex syntactical and organizational
arrangements are possible when thought can be stabilized by
writing.[14]

So what happens when one engages hypertext, which creates neither the visual implication of a coherent subject of the imaginary
order nor the verbal implication of a unitary subject of the symbolic
order? The Internet is a decentered field. It does not position the
subject visually, as Renaissance painting does, nor does it linguisti-
cally "hail" the reader, as classic narrative does. Individual voices
exist, but they are embedded within a heterogeneous and boundless
"text" of many voices. Individual images may employ perspectival
illusionism, but those images are embedded within a vast collage of
images conforming to every conceivable visual convention. No unified narrator or spatially consistent image brings coherence into the
Internet as a whole. Rather, such voices and images are absorbed
into a larger heterogeneity. The subject is neither fixed spatially, as
it would be when looking at a perspectivally correct, illusionistic
painting, nor is the subject addressed as a subject, the way it is addressed when entering the symbolic, when hailed in language and
culture. Thus, hypertext tends to push further the modernist tendency to disperse and decenter the subject.

On the Internet it is easy to become disoriented. Although one is free to navigate as one wishes, to choose this path or that, as one pleases, this apparent freedom is largely specious. Because there is no map of the Internet, the choices one makes often are whimsical rather than significant. One chooses, but on the basis of insufficient information. Although various tools are being developed that will enable us to make navigating the Internet more coherent and informed, at present it is a little like being dropped in the woods without a map or compass and being told we are free to choose whatever path we like.

Moreover, on the Internet individuals can remain anonymous or build "selves" that have little to do with the selves they are in the nonvirtual world. Internet encounters can occur between individuals who know little about one another, and creating alternate identities is always a possibility. Individuals we encounter on the Internet may know something about our professions or interests, because of the list we are members of or the Web site we visit, but, unless we reveal such information, they don't know what we look like, our race, our class, or what kind of neighborhood we live in, and they don't even have to know our gender. Moreover, they can't hear the rhythms, pitch changes, and dynamic shifts that occur when we use our voice; all these important aspects of our communication and our identity, of the persona that we present in the nonvirtual world, are unavailable. Thus, on the Internet there is a flattening and abstraction of the personality and the possibility of compensating for that abstraction with other details that may or may not conform to the individuals we are outside the Web. It may be that the personality is always a mask, but on the Internet the wearing of the mask is foregrounded. This fact, combined with the disorientation one experiences in navigating a labyrinth with so many unmarked paths, has the effect of emptying the identity or, if not emptying it, underlining the fact that it is a construction.

The disorientation one can feel on the Internet—the labyrinthine problem of finding and losing one's bearings—combined with the fact that we often cannot be certain about the identities of those we encounter, can contribute to a feeling of anxiety and paranoia. Moreover, although we may feel anonymous on the Internet, we are also aware of a contrary possibility: that we are being watched. Hidden marketers or investigators or hackers may be tracking our paths, the sites we visit may be documented, bank account and credit card numbers may be stolen. We may try to hide, but we cannot be sure that people don't know more about us than we want them to know. The effect is like Jeremy Bentham's panopticon, which figures so significantly for Michel Foucault. The panopticon is a clever and efficient surveillance structure built on the principle of uncertainty regarding whether or not any particular individual is being watched at any particular moment. That uncertainty ensures the fact that the (possibly) observed individual will behave, even when not actually being watched.[15]

The ultimate technological paranoia occurs when we imagine that cyberspace may, in fact, contain entities or be itself an entity that, like cyborgs, develop motives and wills of their own, at odds with those of their human makers. Novels like *Neuromancer* and films like *Ghost in the Shell* contain examples of such entities—Wintermute and Neuromancer in *Neuromancer;* the "Puppet Master" in *Ghost in the Shell.* Such entities present themselves as hostile forces, though in the end, as is the case in these instances, they may reveal themselves to be quasi-religious higher beings, the next stage in an evolutionary process toward a higher form of consciousness.

It is also important to consider the way in which the Internet, as well as other kinds of hypertext, is divided up, articulated. The Internet, as a field of visual and verbal representation, combines characteristics that create apparently clear articulations of its semiotic field with characteristics that subvert those articulations. Articulations or lack of articulations in visual or auditory fields can be related metaphorically to the sense of the body and its separation or nonseparation from that which is outside the body. Thus, even nonobjective works of visual art can be seen as metaphorical statements about the body, the way it is articulated, both within itself and between itself and the world, or the way in which its parts spill amorphously one into the other and into the world. Any semiotic field, even an abstract one, can be seen as a figure for the organization of the body and its separation from the world or for the disorganization of the body and its lack of separation from the world. If articulations or subversions of articulations in visual and verbal fields are metaphors for the way in which the body and the world are articulated, and if engaging visual and verbal fields involves a mirroring by the subject of those visual or verbal fields, then what happens when the subject engages hypertext? Is the image in the glass coherent, unified, organized, and separate, or incoherent, disunified, disorganized, and permeable?

The individual page in hypertext—what George Landow calls a

lexia, using the term that Roland Barthes uses in *S/Z* to refer to a segment of text—seems to be clearly articulated, a section of hypertext that, though linked with other pages, has its own unity and integrity.[16] However, within the same page, as it appears on an individual screen, other pages or images imported from other pages may be embedded. A screen that seems to be a unity may in fact be a construction of disparate screens from separate sources. Thus the "skin" of a page is permeable; it can be crossed, penetrated, transgressed. At this point, such a construction is deconstructed at the outset because of the disparity in time that exists as text, background, and images are accessed. Although there is little doubt that this problem will soon be solved and the speed of linking to a complicated page will become seamless and the constructed nature of the pages less evident, at present the heterogeneous nature of a page is foregrounded.

Moreover, on the World Wide Web, individual pages link to the pages of other Web sites almost as easily as they link to pages from the same Web site so that the sense of separation of one part of the net from another part is all but lost; when one navigates the Web one slips messily from one page to another, from one site to another, losing coherence, losing a sense of center, losing a sense of the shape of the path one has taken, and losing a sense of where one site begins and another site ends. Landow speaks of "atomization" and "dispersal" in hypertext: atomization because of the small individual unit, but dispersal because of the sense of formlessness and decenteredness that results when one links from one segment to another:

> At the same time that the individual hypertext lexia has looser, or less determining, bonds to other lexias from the *same work* (to use a terminology that now threatens to become obsolete), it also associates itself with text created by other authors. In fact, it associates with whatever text links to it, thereby dissolving notions of the intellectual separation of one text from others.[17]

That is, the "body" of a hypertextual work—whether that work is a conventional work placed on the Internet, a Web site, or a literary work created specifically as a hypertext—has a weak, highly permeable skin and, indeed, may come to be experienced as having no skin at all. The distinction between parts of itself—pages or lexias that are part of it—and parts of other works or sites is weakened; the separation of the pages of one site and those of another to which it is linked becomes nearly indistinguishable from the separation of pages within the site itself. Thus hypertext challenges the very idea of a "work" as a unified text or representation, separated from other texts or representations. And, if we regard any text as a metaphor for the subject of the individual who engages, who mirrors it, then such hypertexts imply a weakening of the distinction between self and other.

Although the cyborg and the net recapitulate an ancient binarism—the material or carnal and the mental or "spiritual" (as they manifest themselves in Molly and Case in *Neuromancer*, for example)—together they suggest similar anxieties and hopes we feel as we move into the highly technologized future. At first blush, we are faced with the hope of enhanced power and freedom. Without such a hope, individuals would be little motivated to participate in this hypertechnologization. Implants and protheses supplement our physical and perceptual abilities or compensate for deficiencies or losses we may experience. The Internet supplements our mental and communicational capabilities, creating a huge electronic system of communication, information, and information-processing that emerges almost as an entity itself.

The image of the cyborg, as well as that of a related figure, the replicant, challenges the conventional distinction between human and machine, and, largely because of the work of Haraway, the cyborg emerges as an important metaphor for other boundary crossings as well—those of race and gender most significantly. Such hybrids challenge conventional categories and make us aware of the fact that all categories are conventions. Indeed, what counts as a "hybrid" is itself arbitrary because it is dependent on conventionally established nonhybrid categories. Under different conventions, hybrids might be seen as "pure" and what we now view as "pure" might be seen as hybrid.

But the image of the cyborg also suggests the possibility of violation, the human skin penetrated and invaded by a technological Other that is alien. And cyberspace, though it seems to enlarge our possibilities for communication and access to information, similarly presents us with the possibility of violation. We experience the freedom that anonymity provides and the possibility for testing new identities. At the same time, we may not be as anonymous as we think, and our movements may be watched and recorded by invisible observers who have purposes different from ours. Paranoia, suspicion, and the fear of surveillance result.

Moreover, on the Internet there exist contradictory possibilities of empowerment and diminution. We can be empowered and enlarged because of the communicational possibilities and free access to information, but we can be diminished within the large ongoing conversation that we become a small part of. And, to the extent that we mirror the media we engage, we ourselves become decentered and dispersed when we engage the Internet, which lacks both visual and verbal coherence and focus and presents to us a field consisting of multitudinous, penetrable elements that spill amorphously one into the other. Like the cyborg, these elements can be assembled and reassembled in many ways, creating possibilities for terror and bliss alike.

Notes

1. Donna Haraway, "A Manifesto for Cyborgs: Science, Technology, and Socialist Feminism in the 1980s," *Socialist Review* 80 (March/April 1985): 93, 68–70.

2. See essays by DeeDee Halleck, "Watch Out, Dick Tracy! Popular Video in the Wake of the *Exxon Valdez*," in *Technoculture*, ed. Constance Penley and Andrew Ross (Minneapolis: University of Minnesota Press, 1991), 211–29; Andrew Ross, "Hacking Away at the Counterculture," ibid., 107–34; and Houston Baker, "Hybridity, the Rap Race, and Pedagogy for the 1990s," ibid., 197–209.

3. Donna Haraway, "The Actors Are Cyborg, Nature Is Coyote, and the Geography Is Elsewhere: Postscript to 'Cyborgs at Large,'" in *Technoculture*, 24–26.

4. Haraway, "Manifesto," 72, 82.

5. "Cyborgs at Large: Interview with Donna Haraway," in *Technoculture*, 9.

6. William Gibson, *Neuromancer* (New York: Ace Books, 1984), 25.

7. See Julia Kristeva, *Powers of Horror: An Essay on Abjection*, trans. Leon S. Roudiez (New York: Columbia University Press, 1982).

8. Kathy Acker, *Empire of the Senseless* (New York: Grove Weidenfeld, 1988), 3.

9. Jacques Lacan, *Écrits: A Selection*, trans. Alan Sheridan (New York: Norton, 1977), 1–7.

10. Louis Althusser, *Lenin and Philosophy and Other Essays*, trans. Ben Brewster (London: New Left Books, 1971), 162–66.

11. See Catherine Belsey, *Critical Practice* (New York: Methuen, 1980), 56–70.

12. Lacan, *Écrits*, 65–68.

13. Walter Ong, *Orality and Literacy: The Technologizing of the Word* (London: Methuen, 1982), 178.

14. Ibid., 37–39.

15. Michel Foucault, from *Discipline and Punish* (1977), in Anthony Easthope and Kate McGowan, eds., *A Critical and Cultural Theory Reader* (Toronto: University of Toronto Press, 1992), 84–87.

16. See George Landow, *Hypertext: The Convergence of Contemporary Critical Theory and Technology* (Baltimore, Md.: Johns Hopkins University Press, 1992), 4 and passim; also Roland Barthes, *S/Z,* trans. Richard Miller (New York: Hill & Wang, 1974), 13.

17. Landow, *Hypertext,* 52–53.

Shelf Life

Geoffrey Bennington

Emory University

"There's dying, then there's dying."
—William Gibson, *Mona Lisa Overdrive*

H ERE'S a loose, provisional, hypothesis to start, which I advance with no particular terminological caution: namely that there's a difference to be thought about between novels which represent or project a world in which novels themselves are a significant feature, and novels which do not. I'd like to suggest that so-called realist novels have as an essential feature that they naturalize themselves, the fact of their own existence, by referring the reader to a world *in* which novels habitually appear or might easily appear. There's nothing particularly odd about reading a novel, the thought would go, so long as that novel gives rise to a world in which there's nothing particularly odd about reading a novel. The most realist novels, on this view, would be fundamentally "reflexive" or "self-referential," to the extent that they include themselves, virtually at least, in the world they are also in some sense representing or creating. In a certain sense, the *most* realist moment of a realist novel is the moment when a character is described reading a novel, or, by extension, some other printed text. Naturally, and familiarly enough, this moment, based as it is on a specular arrangement, is *also* potentially a moment of crisis for realism: let scenes of reading become important in a realist novel, and they will generate uncanny effects, *mise-en-abyme*, and a sort of hyperrealism that at least since Flaubert (and in fact since Cervantes)[1] can give rise to what we can loosely call a deconstructive moment in so-called realist fiction. The principle of realism would then be the same as the principle of its exasperation and deconstruction—and I imagine that this type of analysis might be extended to other representational media, such as cinema.

Which would mean that something different happens when we are faced with a novel in which novel-reading (or text-reading more generally) does not, and in principle would not, occur and perhaps more especially when its nonoccurrence (or rarity of occurrence) is marked in some way. This is the case, or so it seems to me, in the novels of William Gibson,[2] where there is no shortage of reference to books or other print, but where these are systematically marked as old, out-of-date, or out-of use. So, for example, in *Count Zero* the recurring character called "the Finn" (to whom, or to which, I shall return at length) lives in a labyrinthine dwelling full of ageing and broken objects, including books:

> He looked to Bobby as though he could survive on a diet of mouldering carpet, or burrow patiently through the brown wood pulp of the damp-swollen books stacked shoulder-high on either side of the tunnel where they stood . . . Lucas extended his cane and prodded delicately at a dangerous-looking overhang of crumbling paperbacks . . . "Don't fuck with those first editions, Lucas. You bring 'em down, you pay for 'em."[3]

In a fine essay on Gibson, David Wills refers perhaps a little hastily to "the absence or near-absence of print media in Gibson's science fictions."[4] But there are in fact a number of significant references across the novels, all, however, signifying the out-datedness or arcane nature of print and thereby confirming one of Wills's more general points about the place of "history" and anachronism in Gibson's work. Later in *Count Zero*, for example, a character uses a dictionary to prop open a window. In *Neuromancer* there is a defamiliarized description of a library in the Tessier-Ashpool villa Straylight: "There had been a room filled with shelves of books, a million flat leaves of yellowing paper pressed between bindings of cloth or leather, the shelves marked at intervals by labels that followed a code of letters and numbers."[5] In *Mona Lisa Overdrive*, the eccentrically learned and driven character Gentry, in obsessional search of the ultimate Shape of the matrix of cyberspace, has his eccentricity marked by the fact that, along with his more or less exotic electronic gear he possesses books, a sort of library of Alchemist's lore: "and books, old books with covers made of cloth glued over cardboard. Slick had never known how heavy books were. They had a sad smell, old books." . . . "the sagging shelves stuffed with ragged, faded books" . . . "Gentry pacing back and forth in front of his books, running the tip of his finger along them like he was looking for a special one."[6] In the short story "Burning Chrome" (1986), a sort of anticipatory version of much of what is to come in the novels, the Finn's place is described with "scrap waist-high, inside, drifts of it rising to meet walls that are barely visible beyond nameless junk, behind sagging pressboard shelves stacked with old skin magazines and yellow-spined years of *National Geographic.*"[7] And most significantly of all, perhaps, in *Count Zero*, Bobby talks about learning some basic material about the matrix at school, including "how to access stuff from the print library"; asked by Lucas if he ever did that, "'I don't read too well,' Bobby said. 'I mean, I can, but it's work. But yeah, I did. I looked at some real old books on the matrix and stuff'" (*CZ*, 118).

This marked appearance of books as anachronistic, as decaying objects on the verge of being mere rubbish (and therefore potentially very valuable too),[8] is part of a general problematic of rubbish, detritus, usually referred to by the Japanese word *gomi*, that pervades Gibson's work. In *Mona Lisa Overdrive*, this thematic of *gomi* provides the basis for a slightly unclear opposition between England and Japan: compare "This was nothing like Tokyo, where the past, all that remained of it, was nurtured with a nervous care. History there had become a quantity, a rare thing, parcelled out by government and preserved by law and corporate funding. Here it seemed the very fabric of things, as if the city were a single growth of stone and brick, uncounted strata of message and meaning, age upon age, generated over the centuries to the dictates of some now all but unreadable DNA of commerce and empire" (11) with "Portobello . . . With Kumiko firmly in tow, Sally began to work her way along the pavement, past folding steel tables spread with torn velvet curtains and thousands of objects made of silver and crystal, brass and china. Kumiko stared as Sally drew her past arrays of Coronation plate and jowled Churchill teapots. 'This is *gomi*,' Kumiko ventured, when they paused at an intersection. Rubbish. In Tokyo, worn and useless things were landfill. Sally grinned wolfishly. 'This is England. *Gomi*'s a major natural resource. *Gomi* and talent'" (43). Think too of the figure of Rubin, artist as "master of junk," *Gomi no sensei*, in "The Winter Market" (*BC*, 141), and the question "Where does the *gomi* stop and the world begin?"[9]

Books here, then, are at the end of their shelf life, old and rotting, sinking into the piles of rubbish, or else objects of an antiquarian or aesthetic interest and collection. Books figure in this ghostly way in Gibson only as a way of emphasizing that, by and large, books are

finished. Wills has produced incisive analyses of the two most obvious replacements Gibson's universe provides for a culture of print:
on the one hand "simstim" (simulated stimuli), and on the other
"jacking in" to the matrix of cyberspace, this latter worked with the
further distinction between two apparently distinct modes of experience, the one as a sort of abstract geometrical representation of
data,[10] the other as a sort of hyperrealist simulation of the world. In
this respect, namely that books are finished, *all* these texts have a
built-in spectrality we cannot ignore: we witness in printed form a
world where the printed form is in decay, where no one ever reads
a novel. If all books (all writing) involve an irreducible element of
spectrality (as Derrida has shown from his earliest work), these novels mark that fact more explicitly than most, ghosting in this way a
complicated play of past and future which, I imagine, defines science fiction as such.[11] Reading Gibson, it is as if we were already
dead, witnesses to the world of our death, always too soon for simstim or jacking in to either form of cyberspace, just because we are
reading Gibson and not getting the adventure as simstim or by jacking in. It seems to me that Wills is right to suggest that there is at the
very least a tension between these two ways of representing cyberspace (the geometrical and the hyperreal), and I want to try to develop this a little more (or at least in a different direction) than he
does, by pursuing further the often explicit thematic of *ghosts* in
these texts. If, as the recurring character Molly says in the phrase
I've used as an epigraph, "There's dying, then there's dying," meaning in context not, as might be supposed, that there are different
ways of dying, different experiences of the process leading to death,
but that there are in fact different sorts of death, different ontological characteristics, let's say for now, attached to the state of "being"
dead, then we might assume that it would follow that there would be
different sorts of ghosts, different spectralities affecting, and indeed
defining, those different sorts of death. In Gibson, there are ghosts,
and then there are ghosts. And if, as Derrida argues in *Specters of
Marx*, effects of spectrality are powerful enough to suspend the decisions and demands of ontology in the name of a "hauntology"
where neat distinctions between the real and the spectral, and therefore also of the living and the dead, are difficult to maintain, then
we might expect, and will indeed find, that Gibson's novels in their
own way *also* operate a suspension of ontological predicates, and do
so not so much in Heideggerian fashion by developing a thought of
radical finitude and its associated pathos around being-toward-

death, but by opening up at least the hint of a thought of in-finitude which does not, however, simply revert metaphysically to a thought of infinity or eternity. So my claim will be that Gibson's work is worthy of philosophical interest not so much on the side of the simulation of life, virtual life, if you like, but on the side of what we might call virtual death—but that in this it is in some ways simply working out in spectacular fashion a structural feature of the situation of the reader of fiction in general.

It seems to follow quite naturally that Gibson's work should be full of ghosts, and overtly so. *Mona Lisa Overdrive*, at any rate, opens explicitly and insistently with a ghost or with ghosts, and this thematic will, I want to suggest, dominate the novel as a whole (and by extension the trilogy of which this is the last volume):

> The ghost was her father's parting gift, presented by a black-clad secretary in a departure lounge at Narita.
> For the first two hours of the flight to London it lay forgotten in her purse, a smooth dark oblong, one side impressed with the ubiquitous Maas-Neotak logo, the other gently curved to fit the user's palm. . . .
> Ghosts, she thought later, somewhere over Germany, staring at the upholstery of the seat beside her. How well her father treated his ghosts.
> There were ghosts beyond the window too, ghosts in the stratosphere of Europe's winter, partial images that began to form if she let her eyes drift out of focus. Her mother in Ueno Park, face fragile in September sunlight. (*MLO*, 7)

There are at least three sorts of ghosts elliptically introduced here: 1) the parting *gift* from the powerful father, which, as we shall see, is an agent of protection, but also of control and surveillance (so that the gift is explicitly ambiguous, a parting gift to mark and reduce the parting), a sort of ghost *of* the father, standing in for or replacing him as guide and mentor, but also *replacing* him in a stronger sense, generating in due course with Kumiko an antipaternal, let's say fraternal, complicity; 2) the father's ghost*s* in the plural, which will turn out to be ghosts *of the father* in another sense, ghosts of the father's fathers, not his agents but his mentors; 3) Kumiko's "own" ghosts, on the side of the mother, this time, memories of her mother we soon learn is dead, by a father-abetted suicide, and who will return in a more concrete ghostly form toward the end of the novel.

First, then, the ghost as gift from the father, a gift to accompany his daughter as she leaves to be a *guest* with her father's agent in London. Here he comes:

The ghost woke to Kumiko's touch as they began their descent into Heathrow. The fifty-first generation of Maas-Neotek biochips conjured up an indistinct figure on the seat beside her, a boy out of some faded hunting print, legs crossed casually in tan breeches and riding boots. "Hullo," the ghost said.

Kumiko blinked, opened her hand. The boy flickered and was gone. She looked down at the smooth little unit in her palm and slowly closed her fingers.

"'Lo again," he said, "Name's Colin. Yours?" (*MLO*, 9)

Colin is, of course, a sort of illusion, a mere appearance, a curious sort of *phenomenon* visible and audible only to the person holding the unit, a sort of *virtual* hologram[12] who connects somehow directly to the optic and auditory nerves of the owner, and who "hears" in return the subject's subvocal speech. This radically "private" experience of the ghost is later in the novel broken open when the case of the unit is damaged, and the character called Tick manages to wire Colin into the shared world (what Gibson repeatedly calls the "consensual hallucination") of cyberspace.

It's worth pausing for a moment to follow Gibson's "definitions" of cyberspace or "the matrix" through his work, from "Burning Chrome" ("The matrix is an abstract representation of the relationships between data systems . . . the colorless nonspace of the simulation matrix, the electronic consensus-hallucination that facilitates handling an exchange of massive quantities of data . . . mankind's extended nervous system . . . the crowded matrix, monochrome nonspace where the only stars are dense concentrations of information, and high above it all burn corporate galaxies and the cold spiral arms of military systems," 196–97)), through *Neuromancer* ("a custom cyberspace deck that projected his disembodied consciousness into the consensual hallucination that was the matrix," 12); "'Cyberspace . . . A graphic representation of data abstracted from the banks of every computer in the human system. Unthinkable complexity. Lines of light ranged in the nonspace of the mind, clusters and constellations of data,'" 67), to *Count Zero*: "the infinite reaches of that space that wasn't space, mankind's unthinkably complex consensual hallucination, the matrix cyberspace, where the great corporate hotcores burned like neon novas, data so dense you suffered sensory overload if you tried to apprehend more than the merest outline," 62); "'Sure, it's just a tailored hallucination we all agreed to have, cyberspace, but anybody who jacks in knows, fucking

knows it's a whole universe,'" 170). This last instance has the Finn
(the speaker here) referring to "things out there. Ghosts, voices."
Wills suggests that by *Mona Lisa Overdrive* this quasi-geometrical representation of the matrix has been, as it were, so overtaken by the
"ghosts" that it is experienced no longer as a matrix but as a hyperreal environment, but this is not strictly accurate: "People jacked in
so they could hustle. Put the trodes on and they were out there, all
the data in the world stacked up like one big neon city, so you could
cruise around and have a kind of grip on it, visually anyway, because
if you didn't, it was too complicated, trying to find your way to a
particular piece of data you needed"(*MLO*, 22); "Slick didn't think
cyberspace was anything like the universe anyway; it was just a way of
representing data. The Fission Authority had always looked like a
big red Aztec pyramid, but it didn't *have* to; if the FA wanted it to,
they could have it look like anything" (*MLO*, 84; see also 218 and
278); "A cubical holo-display blinked on above the projector: the
neon gridlines of cyberspace, ranged with the bright shapes, both
simple and complex, that represented vast accumulations of stored
data . . . The colored cubes, spheres and pyramids" (*MLO*, 254; see
also 268 and 270–72). Two recurring features of this description are
worthy of interest: 1) it is quite unclear what value to place on the
motif of *consensus* here, and how to understand the idea that the matrix is the result of a universal *agreement*, and more importantly 2)
the insistence on the placelessness of the matrix, its nonspace, non-
locality, allowing for the slogan "There's no there, there," which
gives its title to chapter 7 of *Mona Lisa Overdrive* and is repeated twice
therein, once with the comment "they taught that to children, ex-
plaining cyberspace" (*MLO*, 55: cf. *Idoru*, "I know you've come from
there, but its there . . . isn't there!" 233). The thought that data
unmanageably large for linear grasp might better respond to some
form of pattern recognition is pushed further in *Idoru*, where Laney's talent, which he is quite unable to understand or control, consists in his ability to *see* "nodal points" in vast fields of data unrespon-
sive to normal analysis. It is worth pointing out the similarity
between Laney's technique for spotting such points and the "freely-
floating attention" of the analyst, as recommended by Freud.

Unlike the experience of simstim, in which the subject is the passive recipient of the sensory experiences of the "agent" (which is
usually, though not necessarily, a human agent, but is a sentient
creature at least),[13] let's say that one of the defining features of the
ghost as described here in the shape of Colin is that it *responds*: it

interacts with the subject in a way which goes beyond, say, mere information retrieval, even though Colin is really a modified version of what was designed as a tourist guide. When Kumiko asks him "What are you?" we get the following exchange:

> "A Maas-Neotek biochip personality-base programmed to aid and advise the Japanese visitor in the United Kingdom." He winked at her.
> "Why did you wink?"
> "Why d'you think?"
> "Answer the question!" Her voice loud in the mirrored room.
> The ghost touched his lips with a slim forefinger. "I'm something else as well, yes. I do display a bit too much initiative for a mere guide program. Though the model I'm based on is top of the line, extremely sophisticated. I can't tell you exactly what I am, though, because I don't know." (*MLO*, 204)

This initiative, which is, however, strictly limited by its reliance on the subject's switching the unit on to summon up the ghost (a page later Colin says "I'm only 'here' when you activate me"),[14] is captured strikingly in *Idoru* (apparently set in a much nearer future), when the hero, Laney, is faced by the holographic star of the novel's title: "He looked into her eyes. What sort of computing power did it take to create something like this, something that looked back at you?" (*I*, 237; see also 238: "But you're just information yourself, Laney thought, looking at her. Lots of it, running through God knows how may machines. But the dark eyes looked back at him, filled with something for all the world like hope").[15]

In other words, if we can transpose this into a Levinasian register, what sort of computing power does it take for a ghost to have a face? Or, to telescope brutally a set of suspicions we are slowly building up, is it possible, through Gibson, to envisage that something clearly not of the order of *Dasein* (something that's only "here" when activated, but that in an important sense dwells there where there is no there)[16] can enter into an ethical relation, marked here by its having a face? (In *Idoru*, Gibson's explicit reference is more Deleuzian than Levinasian, but this may be due essentially to an equivocation over "desire."[17] From Levinas's perpective, the face is where exteriority, transcendence and the infinite shine through the phenomenon: this means that Laney's *quantitative* question ("How much power?") hides what is essentially a *qualitative* shift. Once the construct has a face, the computing power, however "impressive," is no longer the issue.

This possibility is made systematically the more striking by Gibson because of the way he stresses negatively certain aspects of the *hardware* involved in producing this possibility of the face. The point about the hardware in each case is that it is clearly marked as being faceless, or more exactly, *featureless*. These constructs have nothing as familiar as a monitor or a screen. Thus, in *Mona Lisa Overdrive*, the slab of "biosoft" into which Bobby is plugged, and whose abyssal insertion into the matrix generates the biggest core of data anyone has ever seen, is repeatedly described in these terms: it's a "fat gray package"; a "featureless gray package"; a "featureless, unmarked unit" and a "flat, gray package"; a "featureless slab"; a "slab"; a "gray slab"; a "gray box"; a "solid rectangular mass"; a "gray box"; it's "heavy, like trying to carry a small engine block."[18] Similarly, in *Neuromancer*, the "construct" of McCoy Pauley is described as "resembl[ing] the magazine of a large assault rifle" (85).

These two featureless boxes are not quite the same sort of thing, however, and I shall return in a moment to the biosoft in *Mona Lisa Overdrive*, which reserves a few surprises for us yet. But the box in *Neuromancer* is more of the order of Kumiko's ghost, Colin, and part of a series of such "constructs" in Gibson. The essential difference between Colin and the *Neuromancer* construct, at least as far as our concern with ghosts is concerned, is not that the one appears as a virtual hologram and the other has to speak through a computer interface, but that the one (Colin) is of the nature of a fictional character ("What you are is some Jap designer's idea of an Englishman!" says one of the other characters when he encounters him (269)), whereas the other is, well, the ghost of a dead man, a "personality construct" that is basically a sort of recording of the subject which can then continue to respond beyond the biological death of that subject.

This is a version of shelf life because, in *Mona Lisa Overdrive*, Kumiko's powerful Japanese industrialist father keeps a row of boxes on a shelf:

four of them, black laquered cubes arranged along a low shelf of pine. Above each cube hung a formal portrait. The portraits were monochrome photographs of men in dark suits and ties, four very sober gentlemen whose lapels were decorated with small metal emblems of the kind her father sometimes wore. Though her mother had told her that the cubes contained ghosts, the ghosts of her father's evil ancestors, Kumiko found them more fascinating than frightening. If they did contain ghosts, she reasoned, they would be quite small, as the cubes them-selves were scarcely large enough to contain a child's head.

Her father sometimes meditated before the cubes, kneeling on the bare tatami in an attitude that connoted profound respect. She had seen him in this position many times, but she was ten before she heard him address the cubes. And one had answered. The question had meant nothing to her, the answer less, but the calm tone of the ghost's reply had frozen her where she crouched, behind a door of paper, and her father had laughed to find her there; rather than scolding her, he'd ex-plained that the cubes housed the recorded personalities of former ex-ecutives, corporate directors. Their souls, she'd asked. No, he'd said, and smiled, then added that the distinction was a subtle one. "They are not conscious. They respond, when questioned, in a manner approximating the response of the subject. If they are ghosts, then holograms are ghosts." (173–74)

This, it seems to me, is where Gibson's ghosts really start being inter-esting. Let us carefully distinguish these ghosts from others. In *Count Zero*, for example, the billionaire Virek appears across a number of more or less reliable incarnations: his ill body is in what Gibson calls a vat, in Stockholm, but Virek's wealth allows an elaborate set of structures to control his "appearances": "Señor is wealthy. Señor enjoys any number of means of manifestation" (152), says Paco, Vir-ek's amanuensis whom Virek himself (or one of his manifestations) also describes as a "sub-program" (28). But for all the complexity of Virek's existence (so that he is not quite sure of everything his vari-ous manifestations are up to),[19] we are led to think that it is still in some meaningful sense his "living" mind that speaks through its manifestations to Marly. And when, at the end of the novel, Virek is reported to have really died, the manifestations cease and the com-plex business empire Virek controlled breaks up. Virek, then, is a ghost in only quite a weak sense, still answerable, in however compli-cated a way, to the basic ontological distinction of being and nonbe-ing, life and death. First Virek is alive, only with the help of complex prostheses, it is true, but alive nonetheless. The manifestations re-main secondary representatives of something else. And this is true too, in however disconcerting a way, with the manifestations of Arti-ficial Intelligences. In *Neuromancer*, for example, the AI called Win-termute appears to Case in a variety of forms, including that of the Finn himself, and in *Mona Lisa Overdrive*, Kumiko's mother seems to show up in cyberspace, only to turn out to be a manifestation of something else, as we shall see. This is no longer the case with

McCoy Pauley, who *is* dead, or at least who *has died*, with the boxes on the shelf in Kumiko's father's office, and especially, again in *Mona Lisa Overdrive*, with the construct of the Finn encountered by Molly (or Sally) and Kumiko. It is in the context of this encounter that Kumiko remembers the boxes in her father's study.

The Finn is arguably the most striking of the constructs.[20] The others, for example, share with a "fictional" construct like Colin the fact of being switched on and off by a user. Colin is designed as a sort of glorified guidebook, and Case in *Neuromancer* switches Pauley on only when he needs his specific skills to pilot his ice-breaking software program, just as Kumiko's father, or so one imagines, switches on his boxes only for the specific purpose of gaining advice. Although this gives rise to some quite complex and even humorous situations,[21] there is here an essential continuity between referring to a reference manual or a piece of more conventional software (an expert system, for example) and accessing these somewhat more interactive constructs. When switched off, for instance, Pauley returns to a state without consciousness or memory and has no temporal experience of the time he spends switched off. To this extent, he is less "aware" than Colin of his ontological status and has to be told by Case that that is what he is. This generates some theoretical complexity (Pauley is a ROM which can only add to the stock of memories already in him a temporary real-time memory of the current session during which he has been switched on: this leads to some minor inconsistencies (he seems to remember after the first time that he is just a construct, for example) and some pathos in that Pauley wants to be erased as a reward for his help, as though the limbo of the construct were not really nothing at all, but a sort of purgatory from which relief would be welcome.[22] The Finn, however, is not so obviously in the same situation: his construct is permanently switched on, so far as one can gather, and insofar as it has some possibility for visual sensory input (Pauley responds to speech and can "see" the matrix, but not Case, for example, who is "working" him), and some volitional control over what he "does": although he sits in an alleyway in a cast-iron protection,[23] he can apparently decide at least some things about his modes of functioning: "Real-time memory if I wanna. wired into c-space if I wanna" and can even manipulate some sort of scanning device (he's able to tell Sally/Molly that Kumiko is "clean . . . no implants" (*N*, 172, 178)) and use some sort of laser weapon for self-defense. In this case, it becomes quite unclear what it means to say that the Finn is alive or dead, and the assurance about the boxes on the shelf that "they are not conscious" would appear difficult to sustain in this case.[24] It looks like here, in fiction at least, "thought can go on without a body," and it looks as though the Finn, dead though he "is," still has a world, even if his interaction with that world is limited (though there appears to be no theoretical reason why such a construct could not be equipped with appropriate robotic means to move and interact physically and even, why not, concernfully with a world).

But at least we still know, in a sense, *where* the Finn is. So far we have on the one hand a cyberspace with no "there," no *Da*, and spatially identifiable constructs with no (living) *Sein*. The *real* complexity of Gibson's work comes from the combination of the two. In other words, dead people returning in the real world as ghostly constructs on the one hand (or virtual constructs or manifestations), and real people in the virtuality of cyberspace are both more or less manageable within the ontological terms of the tradition: but dead people *in cyberspace* look more difficult. Cyberspace is the intersection of more or less elaborate possibilities for "real people" to "move" in ghostlike fashion away from their "physical" *Da* (what *Idoru* calls "telepresence" (1; see also *VL*, 15–16, 20, passim) and for dead people to achieve the same level of "reality."

This appears to be because the combination of the loss of the *Da* and the loss of the *Sein* leaves a much greater degree of uncertainty than in our previous examples as to the ontological status of what we no longer really have a name for. This situation comes to a head in the final chapter of the final book of the trilogy, in which it appears that all the "characters" left are dead (Angie and Bobby, the Finn and 3Jane), "virtual" (Colin), or AI (Continuity). Even within the already quite discomforting terms of this novel, this final chapter is difficult to read: for example, from the vantage of this "place" ("in this France that isn't France," *MLO*, 314)), it is the apparently "human" characters that seem to be of doubtful status, and when there is an apparent manifestation of Kumiko, who for all we know is still "alive," Angie wonders: "Had the girl seen her, returned her gaze?" 313: here it is the "human" who seems not to enter into the face to face).

The key to what is happening here seems to be provided by the status of the featureless grey slab of biosoft that Bobby lies connected to. At the outset, rather like the slab containing the personality construct of McCoy Pauley, this object is not part of the matrix, but what is referred to as "dead storage" (*N*, 84). But the status of Bobby's lump is harder to assess:

> Gentry said that the Count was jacked into what amounted to a mother-huge microsoft; he thought the slab was a single solid lump of biochip. If that were true, the thing's storage capacity was virtually infinite; it would've been unthinkably expensive to manufacture. . . .
>
> "He could have anything in there," Gentry said, pausing to look down at the unconscious face. He spun on his heel and began his pacing again. "A world. Worlds. Any number of personality-constructs . . . If this is an aleph-class biosoft[25] he literally could have anything at all in there. In a sense, he could have an *approximation of eveything*." (*MLO*, 163)

And later, when the slab is plugged into cyberspace, the huge data-form it represents is described as "complete anomaly, utter singular-ity," again a character exclaims "could have a bloody world, in there," and Colin says "This is a wonderfully complex structure. A sort of pocket universe" (*MLO*, 271, 272, 274). "In" this pocket uni-verse within the parallel universe of cyberspace, all types of ghosts we have identified can apparently co-exist and interact on the same level, giving rise to some curious effects. For example, what looks like Kumiko's (dead) mother appears and injures one of the charac-ters: she then, having been unmasked by Colin as really the (dead) lady 3Jane Tessier-Ashpool in disguise, and having failed to "kill" Colin (whatever that might mean—Colin says that they've been "fighting something of a pitched battle, at a different level of the command program" (*MLO*, 275)), she "leaves," and the injured character turns out no longer to be injured, because as Colin says "she was so angry, when she left, that she forgot to save that part of the configuration."[26] Similarly in *Neuromancer*, the AI-manifestation of the eponymous system tells the "real" (but currently jacked in and technically "dead" ["flatlined"] Case about his [dead] ex-lover whom he has encountered in the hyperreal world they are all "in": "Stay. If your woman is a ghost, she doesn't know it. Neither will you," and a little later, "To live here is to live. There is no differ-ence" (*N*, 289, 305).

This radicalization of the ghost to the point at which its different forms collapse ("There is no difference") has several consequences. One, at the level of specularity I began from, is that the reader too is drawn into a realization of the radical power of fiction to make things happen: anything can happen in a story, and that "anything" has always to do with the raising or laying of ghosts and the concomitant spectralization of the reader. But within the fiction, it seems to go along with Gibson's being driven to the thought that this type of involuting and invaginated "cyberspace" would in some sense have to be said to "know itself" for what it is. To achieve "sentience." In *Count Zero*, and to a lesser extent in *Mona Lisa Overdrive*, this is recounted in terms of the event that occurs in *Neuromancer* when Case allows an Artificial Intelligence system to learn things about itself beyond the legal limits enforced by the Turing agency. "Once, there was nothing there, nothing moving on its own, just data and people shuffling it around. Then something happened, and it . . . knew itself" (*N*, 223). This "self-knowledge" is mediated through voodoo characters who visit and "possess" Angie Mitchell, partly in order to tell the story of that possession:

> When the moment came, the bright time, there was absolute unity, one consciousness. But there was the other . . . Only the one has known the other, and the one is no more. In the wake of that knowing, the centre failed; every fragment rushed away. The fragments sought form, each one, as is the nature of such things. In all the signs your kind have stored against the night, in that situation the paradigms of *vodou* proved the most appropriate. (*MLO*, 264)[27]

At the end of *Mona Lisa Overdrive*, this slightly mysterious, and still perhaps too anthropocentric quasi-cosmogony (too anthropocentric in that it tends to be referred to the more or less evil originary intentions of Marie-France Tessier-Ashpool in constructing her two AI systems, and to her cloned descendant 3Jane in wanting them to break from human control and in having the "gray slab" of the Aleph constructed), finds itself referred outwards, slightly differently perhaps, to a more disturbing alterity. Angie is asking about the "why?" of "When it Changed," and the Finn replies: "Ain't a why, lady. More like it's a what. Remember one time Brigitte told you there was this other? Yeah? Well, that's the what, and the what's the why" (315). This refers to the matrix, in its self-knowledge achieved by the unification of its two warring parts, achieving also knowledge of an other matrix, elsewhere in the universe. The apparently achieved identity of the matrix with itself is immediately compromised by this *other* alterity, an alterity beyond the ultimately dialectical rift within the matrix in its original form (Colin explains: "When the matrix attained sentience, it simultaneously became aware of *another matrix,* another sentience"). It follows quite consistently from this that the parameters of the human are exceeded ("nobody's talkin' *human,*" says the Finn); but this nonhumanity here figured in science fiction terms as another place (an alien planet no doubt, in another galaxy, certainly, that the novel wisely does not get into, ending here) is in fact inscribed from the start in the multiple ghostings we have followed. From the start, ghosts are not quite human, the hauntings already exceeding not only the resources of humanism, as we have seen, but even those of the existential analytic.

But if this is true, and if its being true means it is true *from the start,* then, however striking and occasionally lurid Gibson's inventions may be, that truth must already be inscribed in the original anachronism we identified in the situation we find ourselves in reading these novels. The sort of life I have here called shelf life, half-life if you like, that of the construct, is already written into the written form, less dramatically, of course, but as necessarily as in the technological inventions Gibson imagines. As Derrida showed long ago, writing is inherently inhuman, the trace structure laid out in the *Grammatology* already the necessary possibility of ghosting (and of the becoming *gomi* and its retrieval—pulp fiction, if you like) in general. Gibson's disturbing imagination lets us see more easily something intrinsic to writing in the most general sense. Reading is already in excess of the existential analytic: *Dasein* does not read. Reading is already involved with ghosts in a way that has the resources to dedramatize the relation to death that still generates so much pathos in Heidegger. Reading is the death of us all, all of us already ghosts and constructs.

Notes

1. And even since Homer, where this moment occurs, not of course around the moment of reading, but that of listening to the "song" or referring to such a song, or even commenting on a character's narrative skill. I am not really interested in making this formal point the basis for any historicizing.

2. I shall be primarily concerned with the so-called sprawl trilogy, i. e., *Neuromancer* (1984), *Count Zero* (1986), and *Mona Lisa Overdrive* (1988), which use quite traditional Balzacian methods (recurring characters, explicit cross-references, references to events that happen between the novels) to create a fictional world. The later *Virtual Light* (1993) and *Idoru* (1996) also use recurring characters in a rather different world, set less far in the future, with a less strident but perhaps subtler (or more realistic) use of technological possibilities derived from the present.

3. William Gibson, *Count Zero* (London: HarperCollins, 1995), 166–167. Hereafter *CZ*, cited in the text.

4. David Wills, *Prosthesis* (Stanford, Calif.: Stanford University Press, 1995), 83.

5. William Gibson, *Neuromancer* (London: HarperCollins, 1995), 247. Hereafter *N*, cited in the text.

6. William Gibson, *Mona Lisa Overdrive* (London: HarperCollins, 1995), 87, 288, 303. Hereafter *MLO*, cited in the text.

7. William Gibson, *Burning Chrome and Other Stories* (London: HarperCollins, 1995), 19. Hereafter stories from this volume will be cited in the text as *BC*.

8. See Michael Thompson, *Rubbish Theory: The Creation and Destruction of Value* (Oxford: Oxford University Press, 1979). One way of reading Gibson is as proposing a meditation on the complicated relationships between rubbish and "art." This is directly comparable to the "artistic" activities of the AI in the Tessier-Ashpool orbital station, making the artistic boxes that speak so "humanly" to Marly at the end of *Count Zero*. In "The Winter Market," Rubin makes art out of junk, and the fate of the wasting Lise, to which I return below, clearly relates this problematic to the theme of "virtual death" I am following here, though with a sort of humanistic slant that largely disappears in the novels: "It was like she was born to the form, even though the technology that made that form possible hadn't existed when she was born. You see something like that and you wonder how many thousands, maybe millions, of phenomenal artists have died mute, down the centuries, people who could never have been poets or painters or saxophone players, but who had this stuff inside, these psychic waveforms waiting for the circuitry required to tap in" (*Burning Chrome*, 154).

This would have to be compared with the one other example of Artificially Intelligent art, in which a strange approximation of the specularity effect I am associating with realist fiction surfaces, when in *Mona Lisa Overdrive* we learn that the Artifical Intelligence referred to as "Continuity" is writing a book: "Robin Lanier had told her about it. She'd asked what it was about. It wasn't like that, he'd said. It looped back into itself and constantly mutated; Continuity was *always* writing it. She asked why. But Robin had already lost interest; because Continuity was an AI, and AIs did things like that" (59). This "looping back into itself" will become a feature of cyberspace itself and a defining feature of the "virtual death" I will discuss at the end.

9. See also the "Sprawl" more generally, as first described in "Johnny Mnemonic" (*Burning Chrome*) and what *Mona Lisa Overdrive* calls the "rustbelt."

10. Wills, *Prosthesis*, 69–72.

11. In this respect, William Gibson and Bruce Sterling's *The Difference Engine* (London: Gollancz, 1990) would be an interesting case, projecting onto the past of Victorian England a sort of unrealized future possibility.

12. Compare the more "normal" holograms that appear throughout these novels, culminating in the eponymous "idoru" of Gibson's latest novel, *Idoru* (London: Viking, 1996), which are so-called real images, visible in principle to all around, brought out by Gibson in his inclusion of a character who can't in fact see the hologram: "'Blind' Willy Jude . . . [had] been turning his enormous black glasses in the idoru's direction throughout the meal; now he seemed to sense Laney's glance. The black glasses, video units, swung around. 'Man,' Jude said, 'Rozzer's sittin' down there makin' eyes at a big aluminium thermos bottle.' 'You can't see her?' 'Holos are hard, man,' the drummer said, touching his glasses with a fingertip" (179). Hereafter *I*, cited in the text.

13. In *Count Zero*, Bobby describes a collective simstim audience enjoying "some kind of weird jungle fuck tape phased you in and out of these different kinda animals, lotta crazed arboreal action up in the trees" (59).

14. Much later in the book, when Colin seems to achieve a greater degree of autonomy, and after a moment when he is disconcerted to find that various bits of knowledge he should have are not in fact there, he says "I've been designed to advise and protect Kumiko in situations rather more drastic than any envisioned by my original designers. I'm a tactician" (274).

15. There's a similar moment in *Virtual Light* (London: Penguin Books, 1994): the hero, Rydell, sees from his vehicle a hallucinatory figure subsequently revealed to be a hologram of some sort: "Sometimes you saw things up there and couldn't quite be sure you'd seen them or not. One full-moon night Rydell had slung Gunhead around a curve and frozen a naked woman in the headlights, the way a deer'll stop, trembling, on a country road. Just a second she was there, long enough for Rydell to think he'd seen that she either wore silver horns or some kind of hat with an upturned crescent, and that she might've been Japanese, which struck him right then as the weirdest thing about any of it. Then she saw him—he *saw* her see him—and smiled. Then she was gone" (22). Much later, Rydell comes across the "same" woman dancing naked in a bar, only to be told that "she" is a hologram (171–72).

16. This suggestion of a complication of Heidegger's existential analytic might go with a sense that the "as-structure" of phenomena, reserved by Heidegger for *Dasein*, is also troubled by the fiction of what Gibson calls "microsofts" (which appear only in *Count Zero*, I think). Plugging one of these into the socket the characters have behind their ears allows a sort of transient knowledge to become part of one's "personality." Turner uses one of these to know how to fly a plane: "Knowledge lit him like a [*sic*] arcade game, and he surged forward with the plane-ness of the jet" (141). It will be objected that here the "plane-ness of the jet" is nonetheless only available *as such* to a *Dasein*, however "augmented" with technical prostheses, but this is an unconvincing objection, if only because there is no reason to deny access to just that to the jet "itself," which interacts "intelligently" with its pilot so that the distinction between the human and the technical becomes impossible to draw.

17. See the explanations provided by the idoru's creator, Kuwayama: "the result of an array of elaborate constructs we refer to as 'desiring machines' . . . Not in any literal sense . . . but please envision *aggregates of subjective desire*. It was decided that the modular array would ideally constitute an architecture of articulated longing" (*Idoru*, 178). It is obviously beyond the scope of this paper to explore the relationship between Levinas's and Deleuze's concepts of desire, but this paper suggests that, despite appearances, the former ("metaphysical desire," as laid out in *Totality and Infinity*), rather than the latter, is closer to what Gibson can help us to think.

18. In *Mona Lisa Overdrive*, 22, 52, 91, 116, 162, 186, 289, 294, 301 and 302, 306. In *Idoru*, the support for the hologram of the singer is "like a large silver thermos bottle" (236; also 179).

19. When Marly claims to have seen Virek in good health giving a lecture, he struggles to remember, then says, "You saw a double. A hologram perhaps" (*Count Zero*, 25–26).

20. It is no accident that the encounter with the Finn's construct occurs in a chapter called "Ghosts and Empties" or that Sally/Molly should say "Ghosts . . . Lotta ghosts here for me, or anyway there should be" (*Mona Lisa Overdrive*, 170).

21. Apart from the fact of this construct being a way of keeping Pauley "alive" after "death," the problems here are pointed up by a clear affinity between the character and death "in" life: even before he "died," Pauley was already legendary for having "survived" death, and the same happens when Case is "flatlined" more than once in the novel, although in both cases this "death" is temporary and the subject of it then returns to "life" with some experience of the time when he was notionally "dead." In a slightly different way, the character Rez in *Idoru* gets to "watch his own funeral."

22. As in Denis Potter's television play *Cold Lazarus.* At one point in *Neuromancer,* Pauley is referred to as the "Lazarus of cyberspace" (98). The difference between this and what the AI Neuromancer can do is, technically (though this distinction is not really powerful to account for the effects Gibson imagines) that the latter is "full RAM." There is more than a memory of Poe's Mr. Valdemar in the desire to be released from "death" into death in these cases.

23. In his "fire-hydrant," the Finn can "see" but not be seen, thus exhibiting the "visor-effect" Derrida analyzes around the ghost of Hamlet's father, in *Spectres of Marx.*

24. In the story "The Winter Market," this issue is brought out perhaps more forcibly. Lise is dying from some wasting disease and supported by prosthetic means (an "exoskeleton"). Having made lots of money by allowing herself to be in some sense recorded for a form of simstim, she voluntarily dies, in the bodily sense, "becoming" a software construct. In some ways, this early version is more complex than those that appear in the novels, in that the narrator refers to "whatever it is she's since become, or had built in her image, a program that pretends to be Lise to the extent that it believes it's her" (164). This "belief" seems powerful enough to extend to the program's need to continue earning money (as "Lise") in order to pay for its own continued existence (166). The additional complexity here seems to stem from 1) the fact that this "construct" is explicitly a *program* rather than something as solid as a slab of matter which has to be physically stolen or taken, as is the case with the examples we have discussed, and 2) the oddity of saying that the program *believes* itself to be Lise. (Page numbers refer to *Burning Chrome,* cited in n. 7, above.)

25. The reference to Borges's story "El Aleph" is of course not fortuitous.

26. Subsequently, however, the *effect* of the injury persists, like a sort of phantom limb: "'But Colin said she had forgotten . . .' '*I* haven't,' he said, and sucked air between his teeth, working the sling beneath his arm. '*Seemed* to happen, at the time. Lingers a bit . . .' He winced" (*Mona Lisa Overdrive,* 296).

27. In *Count Zero,* it is one of these fragments, the *gomi*-retreiving "artist" AI, which tells the story to Marly: "I came to be, here. Once I was not. Once, for a brilliant time, time without duration, I was everywhere as well . . . But the bright time broke. The mirror was flawed. Now I am only one . . . But I have my song, and you have heard it. I sing with these things that float around me, fragments of the family that funded my birth. There are others, but they will not speak to me. Vain, the scattered fragments of myself, like children. Like men" (311).